# HOW TO MAKE AND USE THE TREADLE IRRIGATION PUMP

CARL BIELENBERG and HUGH ALLEN

Practical Action Publishing Ltd
25 Albert Street, Rugby, CV21 2SD, Warwickshire, UK
www.practicalactionpublishing.com



First published in 1995
Transferred to digital printing in 2008

ISBN 13 Paperback: 9781853393129
ISBN Library Ebook: 9781780444093
Book DOI: http://dx.doi.org/10.3362/9781780444093



A catalogue record for this book is available from the British Library.



Since 1974, Practical Action Publishing has published and disseminated books and information in support of international development work throughout the world. Practical Action Publishing is a trading name of Practical Action Publishing Ltd (Company Reg. No. 1159018), the wholly owned publishing company of Practical Action. Practical Action Publishing trades only in support of its parent charity objectives and any profits are covenanted back to Practical Action (Charity Reg. No. 247257, Group VAT Registration No. 880 9924 76).

Typeset by Dorwyn Ltd, Rowlands Castle, Hants, UK

# Contents

Foreword ... vii

Acknowledgements ... viii

Introduction ... ix

What's so special about the Treadle Pump? ... 1

Assessing the local market for Treadle Pumps ... 6

Cost Issues ... 11

Treadle Pump manufacturing instructions ... 15

1 Making the pump cylinders ... 16
2 Making the pump body ... 22
3 Making the handle and treadle support ... 37
4 Making the piston assembly ... 42
5 Making and installing the valves ... 51
6 Making the wooden parts of the pump ... 54
7 Final assembly of the pump ... 57

Treadle Pump User's Manual ... 60

Appendix A Engineering drawing set ... 67

Appendix B Tool kit components ... 77

# Foreword

The treadle irrigation pump is among the most effective small-scale technological devices to be developed in the last several years. It is able to lift up to seven thousand litres of water per hour using the power of the human body, and can be locally made at low cost[1] in a myriad of small-scale metal-working shops typical of less-developed nations. Its acceptance in Bangladesh where it was first developed in 1984 is extraordinary, with over 500,000 pumps estimated now to be in use.

The current design offered in this manual has evolved from the Bangladesh original into a fully portable pump with both lift and pressure capacity, and is especially appropriate to situations where soils are permeable, and water cannot easily be distributed through canals.

ATI and CARE have developed a package of specialized tools to make the manufacture of the pump simple and efficient. Using these tools they have assisted manufacture and sale of the pump in a number of countries in West Africa, where adoption has been rapid, and the economic impact on enterprises that have bought the pump extremely positive. This manual is an attempt to make this experience more widely appreciated, and to bring to the attention of technicians and development organizations the potential for a broader dissemination of the technology.

The manual is intended to be read primarily by organizations engaged in economic development activities, which have a particular expertise in technologic transfer: it is not intended just as a step-by-step guide for the manufacturer (although it can be used in this way). It is the experience of both the authors that while the production process is easily understood by most artisans, appreciating the details of manufacture and the importance of quality control measures is best assured by the initial supervision of technicians who are trained in the use of the tools illustrated in this manual. It is also our opinion that careful estimation of the market potential for sales of the pump are an important prerequisite to any decision to promote its use: the costs of tooling are too high an investment to be lightly borne by a manufacturer who lacks the assurance of significant sales.

These assurances are most reliably provided as part of a dedicated programme of promotion that leads to the establishment of standardized quality norms, and a critical mass of pump users in a given programme context. It is on this foundation that spontaneous replication of the technology is most likely to occur.

---

[1] At the time of writing, the ex-factory price of the pump in most countries of West Africa where it is made is less than $100.

# Acknowledgements

This manual has been made possible by a collaborative effort between A.T. International (ATI) and CARE International, both of whom have contributed to the technical development of the present model of the treadle pump, and to programmes in East and West Africa that promote its manufacture and use.

Particular thanks are due to:

**Eric Hyman** of ATI for a careful reading of the text, and for contributing substantially to the chapter on cost issues.

**Davidson Njoroge,** whose drawings are, as always, clear and exact, making the text understandable. The manual would have been impossible without this level of quality and attention to detail.

**Ed Perry** at ATI whose pioneering work in Senegal has given us the experience to be confident of what is presented here.

**Chuck and Sue Waterfield** at CARE, and **Susan Swift** of ATI who found the money to pay for the preparation of the drawings and layout.

# Introduction

This manual describes the manufacture and use of a remarkable human-powered pump especially designed for small-scale irrigation. The pump uses the leg muscles in a comfortable walking motion, and can be operated by one or two adults, or two or more children. It can lift water by suction from wells up to 7.5 metres (24.5 feet) deep or from surface sources. Because of its efficient design and effective use of the body's strongest muscles, the treadle pump can lift up to seven thousand litres of water per hour, which is enough to irrigate up to half a hectare of land. The pump is designed to be produced in less developed countries by metal working artisans and small manufacturers, using readily available materials, ordinary workshop equipment, and a low-cost set of specialized tools.

Local artisans have adapted the design of the treadle pump to many different specifications, depending on local farm conditions, raw materials, and their available equipment and technical skills. The version of the treadle pump described in this manual is based on a design developed in Bangladesh in 1981 by Gunnar Barnes, a Norwegian engineer, while working for the Rangpur-Dinajpur Rehabilitation Service. The Bangladesh treadle pump, called by farmers the tapak-tapak pump because of the sound it makes, has been very well received, with well over 500,000 now in use in Bangladesh. This pump is mainly used for lifting water from shallow hand-bored wells for irrigating dry-season crops of rice, wheat and vegetables.

The pump described in this manual and its manufacturing technology were developed under a grant from Appropriate Technology International (ATI), and differs from the original version in two respects.

- While the Bangladesh pump is sold in stripped-down form, and users are required to make and install the handles, treadles, treadle-support and other parts from bamboo and wood, the pump presented here incorporates all these parts into a single transportable unit, and, as such, the pump is ready to use.
- While the Bangladesh pump can lift water only by suction, the water then flowing from the pump to crops through open canals, this pump can lift water from 7.5 metres, and then deliver it under pressure either to additional height above the pump, or up to 100 metres horizontally. The pump has connections for both suction and discharge piping.

This version is more expensive than the original Bangladesh model, but we feel that its added versatility and ready-to-use design will facilitate introduction of the technology into other countries, especially in cases where sandy soils make it impossible to construct irrigation canals. Plans for the original Bangladesh treadle pump may be obtained from the International Rice Research Institute (IRRI).

Over a five-year period ATI has experimented with many variants of the original design, and has settled on this model after assisting small-scale artisans in Senegal to produce and sell over 875 units, on an unsubsidised basis. ATI has also promoted the local manufacture of the pump in North Cameroon, Nigeria and Mali. CARE has introduced the pump into Togo, where early results are promising: with over 100 sales to date the pump is beginning to displace small-scale motor-driven pumps. CARE is also about to initiate production in Zimbabwe.

The concept of a suction/pressure or universal treadle pump was first presented in a practical design by a USAID mechanical engineer named Dan Jenkins in 1985. The Jenkins pump is simple to build, as it is made from standard PVC pressure pipes and fittings. However, such fittings are costly, comparatively fragile, and may not be available in many countries.

The design presented here, like the original Bangladesh pump, is made primarily from mild steel sheet, and is assembled by electric arc welding. It is very durable, if reasonably well maintained, yet is light enough to be easily carried by a single person. The present design represents a trade-off between production cost and complexity on the one hand, and durability on the other.

*It is strongly recommended that prospective manufacturers strictly adhere to this design, for at least the first 50 pumps, utilizing the specialized tools and manufacturing instructions presented in this manual.*

After some time, pump-makers will begin to develop design modifications and adaptations that suit their individual circumstances. It is best, however, to defer these changes until a considerable number of pumps have been in use for at least a single growing season, and users have a chance to provide feedback on performance and problems.

Each pump user should be presented also with a copy of the Treadle Pump User's Manual, translated into local languages. This can be photocopied direct, or obtained in bulk from ATI. Additional copies of the manual, and ready-to-use manufacturing tool kits (illustrated here) and demonstration pumps can also be obtained through:


- Appropriate Technology International
  1828 L. St. NW
  Washington D.C. 20036
  USA

  Fax: 1 202 293 4598
  Tel: 1 202 293 4600

- CARE International
  SEAD Unit
  151 Ellis Street, N.E.
  Atlanta Ga 30303
  USA

  Fax: 1 404 577 1205
  Tel: 1 404 681 2552


The manual is divided into five sections.

- ***What's so special about the Treadle Pump?***

  This explains the unique characteristics of the pump, and, for given applications, its advantages relative to other types of manual and mechanical irrigation pumps.

- ***Assessing the local market for Treadle Pumps***

  This describes for prospective manufacturers and programmers the necessary conditions that must apply for successful introduction of the treadle pump. It also describes strategies employed by ATI and CARE to introduce the pump to a new market.

- ***Cost issues***

- ***Treadle Pump manufacturing instructions***

  This is the heart of the manual, and gives illustrated step-by-step manufacturing instructions.

- ***Treadle Pump User's Manual***

  The Appendices give detailed drawings of the Treadle Pump.


*Carl Bielenberg (ATI)*
*Hugh Allen (CARE)*
*November 1994*

# What's so special about the Treadle Pump?

## 1. General

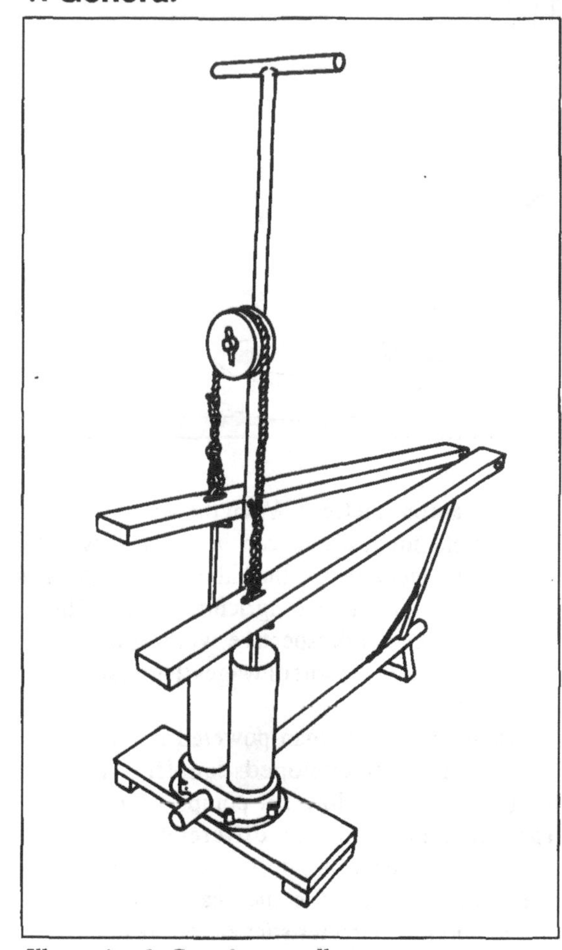

*Illustration 1: Complete treadle pump*

The treadle pump has a number of features that distinguish it from most human-powered pumps. First, it is a high capacity pump designed to lift water from shallow wells and surface sources. Most hand pumps are designed to lift relatively small quantities of water from deep wells and boreholes. Irrigation, even on a small scale, requires large amounts of water, approximately 50 cubic metres (50 000 litres) per hectare per day under average conditions when there is no rain. The treadle pump, with a capacity of 5 000–7 000 litres per hour, can irrigate between 1/4 and 1/2 hectare with a family's own labour. Most hand pumps, which deliver about 1 000 litres per hour, can irrigate no more than 1/10 hectare under average conditions.[2]

Second, the treadle pump lifts water from a well by suction, the pump cylinders that develop this suction forming part of a portable self-contained unit that sits at ground level. Most hand pumps have their drive mechanism and discharge spouts above the ground, but their pump cylinders and pistons are located under water inside the well. This configuration enables the pump to deliver water from great depths, but require that they be permanently installed on a single well. Portability is an important characteristic of pumps used for small-scale irrigation for the following reasons:

- It enables a single pump to be used on a number of wells, each of which (because of low recharge rates) may have the capacity to irrigate only a small area. Many farmers in Africa who irrigate with hand-dug wells have a number of wells on their farms. Multiple wells increase the aggregate quantity of water available, assure a reliable supply, and reduce the distance that water must be physically carried (usually by sprinkler can) to the crop.
- By moving the pump from one location to another, the farmer can minimize the amount of piping or canal that must be available to irrigate the entire farm. This is important because of the very high cost of pipe, and because, on sandy soils, unlined canals lose a large proportion of the water being delivered by the pump: keeping the distance from the pump to the crop as short as possible will minimize these losses.
- The pump is easily dismounted and stored at the end of the day's work, thus reducing the risk of theft or accidental damage.

---

[2] The *Shadoof*, which is common in Egypt and many Sahelian countries, is probably the most cost-effective and efficient traditional water lifting system used for irrigation. Depending on the power of the human arms and upper trunk to operate, it lifts up to 2 500 litres of water per hour from up to four metres depth. The treadle pump gains its distinct advantage because it uses a different muscle group (the legs) which are far more powerful than the human arm, and reduces wasted motion: because it is double-cylindered and double-acting, both down-strokes and up-strokes deliver water.

***Illustration 2 : Typical configuration with two adults***

The treadle pump, like all pumps that lift water by suction, cannot draw from wells that are deeper than 7.5 metres. This is because, when a *perfect* vacuum is created by the pump in its suction pipe, the pressure of the atmosphere pressing down on the water in the well is only sufficient to lift the column in the suction pipe 9.8 metres. If the pump is *imperfect*, and leaks a little air, or if there is a small air leak where the suction pipe is attached to the pump, the maximum lift will be less.[3] Motorized centrifugal pumps, which are the kind most commonly used by affluent farmers in Africa, usually lift water no more than six metres due to cavitation or bubble formation around their fast-turning impellers. A well-made treadle pump can do somewhat better, lifting water by suction approximately 7.5 metres at sea level and 6 metres at 2 000 metres above sea level.

Although the limited lift of the treadle pump makes it inapplicable to areas where the water table lies more than 7.5 metres below the surface, this is not an important limitation for a human-powered pump. This is because the human body has a small and finite amount of power, and thus, as water depth increases, the amount that can be lifted in a given period of time decreases proportionately. A single healthy adult male working on a treadle pump can lift about 6 000 litres of water an hour from a depth of four metres, and approximately 3 000 litres from a depth of 7.5 metres. None the less, the labour cost of lifting one cubic metre of water from 7.5 metres is twice a much as when the water is at four metres. When the water is deeper than eight metres, and can be reached with an alternative technology, the labour cost becomes excessive, and irrigated agriculture unprofitable. Seen from another perspective, workers will not be willing to accept the sort of wages that a farmer can then afford to pay.

There are few human-powered pumps on the market in less-developed countries that are large-capacity shallow lift pumps. Furthermore (and this is a critical difference) there are *none* that are operated by the legs in the natural stepping motion used on the treadle pump. These are important characteristics of the treadle pump, because they allow the operator to work for long periods of time without excessive fatigue. Leg muscles are by far the strongest in the body, and they are harnessed to the pump in the same manner as the legs are accustomed to work, that is, when walking with a load, or climbing a gradual slope. The ability of two adults to work together, facing one another, makes long pumping a tolerable - even sociable - experience. With the hands free one can even read, or dandle an infant while working the pump.

---

[3] In addition, even where there is a perfect vacuum, water cannot be drawn at high volume from a well of more than nine metres depth. Since the mass of water in the suction pipe has considerable inertia, it will allow the development of a vacuum *on top* of the water column if the pistons are cycled at a normal rate, leading to a very much reduced rate of delivery.

## 2. Alternative irrigation pumps to the Treadle Pump

### *The Japy Pump*

One shallow lift hand pump that is widely available in Africa is the *Japy pump*. The Japy is a cast-iron pump that uses a hand-operated lever to rotate a vane in its cylindrical housing. The construction of the Japy is very simple and rugged, and, like the treadle pump, it delivers a continuous stream of water by suction and pressure. The Japy cannot, however, be used for long periods of time, because it is operated by the arms, and not the legs. Conversion of a Japy to foot operation might be possible, although its cost would be substantially greater than that of the treadle pump. For example, the 5m$^3$ per hour model cost 55 750 CFA fr in Bamako, Mali, in 1992 (before devaluation), and the conversion to foot operation was estimated then to add 10 000 CFA fr to this price. Since the Japy is produced by casting and machining, it cannot be produced (or easily repaired) by artisans, and would tend to be expensive if made in small quantities by formal-sector manufacturers who are not suitably equipped to achieve economics of scale.

### *The Rower Pump*

In addition to the treadle pump, there are at least two other pumps that are not in widespread use in Africa, but which may have considerable potential for small-scale irrigation. One of these is the *Rower pump*, a very simple single-acting shallow lift pump that is currently being made in Nepal and Thailand. The rower pump is somewhat less expensive and simpler to produce than the treadle pump, because it has fewer parts, and these are primarily made from PVC. It is an efficient pump, and uses both the arm and back muscles of the human body, and can therefore attain moderately high rates of delivery. Thailand's Appropriate Technology Association, which builds and sells the Rower pump, has tested the standard two-inch diameter pump, and found it can deliver 0.8 litres of water per second from seven metres of depth, while a three-inch pump reportedly can deliver up to two litres per second from four metres of depth, which is equal to the delivery rate of the treadle pump. No information is available as to how long this effort can easily be maintained.[4]

Thus far, the rower pump has mainly been installed in a semi-permanent configuration on shallow tubewells, and hand-dug wells. This is partly because its PVC construction makes it susceptible to breakage and damage due to long-term exposure to sunlight. For these reasons it is usually buried, to provide protection and mechanical support. The pump could, however, be easily installed in a lightweight wood or steel frame, that would securely support the pump, and provide a comfortable seat for the operator. Its intrinsic disadvantages, when compared to the treadle pump are that:

- it does not provide a variable mechanical advantage in the way that the treadles of the treadle pump do
- it cannot be used as a pressure pump
- for the same volume of water pumped it is more tiring to use than the treadle pump.

This means, in the first instance, that the larger 3-inch (75mm) rower pump that is able to deliver as much water per hour as the 4-inch (100mm) treadle pump is strictly limited to no more than 4 metres of lift. The 2-inch (50mm) rower pump, although able to pump from 7 metres, delivers about half as much water as the 4-inch treadle pump. Multiple sizes of the rower pump are therefore indispensable, and must be carefully matched to the depth of water at the site (despite the fact that water levels vary throughout the year), and the strength and weight of the operator. Because the rower pump cannot be used as a pressure pump, it cannot move water through long lengths of pipe, and cannot deliver water to a reservoir located above the pump. These are, however, applications for which only a minority of treadle pumps would be used, but should be considered when assessing the market for each type of pump.

### *The Rope-and-Washer Pump*

The second type of human-powered pump that may be applicable to small-scale irrigation is the *Rope-*

---

[4] In Chad, CARE worked for three years with a version of the three-inch rower pump that was attached to a standard pump lever, rather than directly actuated by the operator. Consistent, continuous delivery of water from four metres depth averaged between 1 800 - 2 500 litres per hour, depending on pump quality, and the vigour of the operator.

*and-Washer pump*. The rope-and-washer pump is not really a pump at all, but lifts water by carrying it between successive washers that move through a pipe, spaced equally apart on a loop of rope. The rope and washer pump is also somewhat easier to produce than the treadle pump, but shares the same disadvantages as the rower pump. While the rope-and-washer pump can be set up to use power from the legs, by attaching pedals to the drive wheel and installing a seat for the operator, it has been generally installed using a hand-powered crank. The hand cranked version quickly tires the operator, and is not well suited to providing the large amounts of water required for irrigation. Usually made of PVC, its construction is also inherently vulnerable to breakage caused by the thrashing motion of the rope on entering the lift tube, and the awkward meshing of the washers with the drive pulley.

The rope-and-washer pump has several limitations. These are that:

- It cannot be used to develop pressure, or lift water above the pump itself.
- It is at its most efficient when the pipe in which the rope travels is vertical, or nearly so.
- It cannot be installed on a tubewell, unless the casing is large enough to accommodate the riser pipe, and the return run of chain.
- The friction of numerous washers travelling in the riser pipe adds significantly to energy losses.

These factors effectively limit its use to hand-dug wells of fairly wide diameter, and preclude pumping from some surface sources such as rivers and small streams with gradually sloping banks.

The rope-and-washer pump can, nevertheless, lift water from greater depths than the treadle pump (or indeed any pump that relies on suction). However, at least two different models of the pump would have to be provided to cover the range of heads (two to six metres) that are comfortably handled by a single adult using the treadle pump.

A third type of human-powered pump that may have limited applicability for small-scale irrigation is the *Diaphragm pump*. This type of pump uses a flexible bellows-like diaphragm, usually made of rubber, which is mounted between two steel plates. By changing the position of these plates up and down, the volume of the water contained in the bellows is altered, thereby creating a pumping action. A number of designs have been developed, one by IRRI that uses a diaphragm of inner-tube rubber, and others that use a worn-out tyre as a diaphragm/water displacement chamber. The diaphragm pump is suited to extremely low lifts, where the head is less than two metres, as may be encountered in the irrigation of lowland rice. It does not suit the majority of circumstances encountered in small-scale irrigation, and for which the treadle pump is well suited.

We have mentioned several alternatives to the treadle pump because each of these pumps is well-designed, and may, under particular circumstances, be more appropriate than the treadle pump. For example, the Japy is certainly the best pump mentioned if the owner's objective is to supply domestic water requirements from a shallow well, or to irrigate a small seedling nursery. The other pumps are not well-adapted to the supply of domestic water, but are for irrigation, for one or more of the following reasons:

- they deliver large amounts of water
- they have open cylinders that may sometimes require priming (which could contaminate the well)
- they prevent sealing of the well.

The treadle pump is slightly more expensive than the rower, rope-and-washer and diaphragm pumps, but is more versatile, and can deliver water through a broader range of heads and for a longer time, at (usually) higher volume, without user fatigue.

## 3. Conclusion

The treadle pump is made from mild steel, and is assembled by arc welding. It can therefore be produced by a large number of metalworking artisans who are usually engaged in making steel gates and window grills, as well as by small- and medium-scale manufacturers of metal products. Although the inside of the cylinders are unpainted, they resist rust because of the lubricating and polishing action of the leather pump cups. The use of sheet steel results in a fairly lightweight pump that is inexpensive, yet able to stand the stresses and wear of long use. Were the pump to be made in a developed country, where unit labour costs are much higher, a treadle pump having an injection-moulded plastic valve box and cylinders of heavy-walled PVC tubing would be much cheaper. The present design is chosen, however, because it is better adapted to the skills, materials and cost endowments that are apt to be found in most developing countries, as well as to the hard conditions of use that are prevalent.

The suction/pressure pump described in this manual is, as already noted, more complex than the original version developed in Bangladesh. Its production is, however, made much easier, and its quality improved by using a kit of specialized tools that have been developed for this purpose. The toolkit can be purchased through ATI or CARE, and costs approximately $1 000. Under typical cost and market conditions prevailing in most of West Africa it can pay for itself in the production of 100 pumps.

Other treadle pump designs, made from cement, PVC, and cast iron have been developed. The diversity of these designs supports our view that this is a practical and important technology for developing countries. None the less, we encourage those who read this manual to understand the rationale behind the present design and put the pump to use before attempting to modify the design. Unique features of the present treadle pump include:

- its versatility, and the ability to deliver water by suction as well as by pressure
- its low clearance, or the volume that exists between the piston and the valves when at the bottom of the stroke. Low clearance helps the pump to prime itself as the suction lift approaches the theoretical limit of 9.8 metres
- attachment of the treadles, via the support tube and intake pipe, to the pump body. This last feature greatly facilitates setting up the pump, carrying the pump, and improves its stability when in use.

The present treadle pump, like the Bangladesh pump, has four-inch (102mm) diameter cylinders. By changing one's position on the treadles, or by working the pump alone without a partner, the pump can be comfortably and efficiently used at between one metre and ten metres of head. However, there are situations in which a larger or smaller diameter cylinder would be more efficient than the present design. In irrigating swampland, or wherever the total lift (including pipe friction) is less than two metres, a larger pump would deliver more water. A pump with five- or six-inch diameter (125 or 150mm) cylinders could be built by scaling up the present design, and would be capable of delivering between ten and 14 cubic metres of water per hour. Experimental pumps built in Togo and Senegal achieved these rates of delivery on low heads.

Where the total lift (suction plus pressure head) is greater than ten metres a pump having smaller cylinders would be less fatiguing to operate, although the practical minimum diameter to accommodate the valves is 80mm, providing a reduction in effort of approximately 35 per cent. The most common situation in which more pressure would be required is in market gardening on sandy soils, where the water is delivered by long pipelines or flexible hose, and delivered to the crop as a spray. The standard size of pump can easily be retro-fitted with smaller cylinders by inserting cylinder liners made from high-pressure PVC tubing, and by installing smaller piston disks and leather pump cups.

# Assessing the local market for Treadle Pumps

## 1. Factors most likely to influence the purchase and use of the Treadle Pump

Efforts by ATI and CARE to introduce the manufacture and use of the treadle pump in Africa have thus far focused on Nigeria, Mali, Nigeria, North Cameroon, Senegal and Togo. These countries have been selected in part because one or both organizations had project activities there, and because a number of factors indicate that substantial demand may either exist or has the potential to be created. This section attempts to identify these factors, and to assist other agencies to identify and assess the local market for treadle pumps.

These factors are:

***Physical environment***

- o The water table is high enough, throughou the dry season, to permit the use of treadle pumps, and the recharge rate of most wells is sufficient to justify their use (one cubic metre per hour).

***Economic environment***

- o Farmers are currently engaged in irrigated agriculture which is:
  - intensive and extensive
  - producing food in significant amounts for local cash markets.
- o Irrigable land is affordable, and easily available to permit the expansion of production.
- o Sufficient labour is available to work in the sector.
- o Sufficient capital is available to invest in improved pumping technology.
- o Farmers have made significant cash investment in fixed assets and working capital. This may include:
  - land purchase
  - motorized pumps, or other mechanical pumping systems
  - concrete reservoirs/basins, permanent underground piping, watering cans, buckets, concrete-lined wells and fencing
  - sizeable wage labour force
  - investment in fertilizer and pesticides.
- o The cost or availability of fuel and/or spare parts is seen as a major constraint on the profitability and expansion of the sector.

***Technical environment***

- o Materials required to make and install the treadle pump are available. These include sheet steel, pipe, round bar, high quality tanned leather, wood, welding rods, nuts, bolts, PVC piping and fittings, polyethylene sheet, inner tubes etc.
- o Local artisans have the equipment and skills required to produce the treadle pump, and are ready to invest working capital in its production and marketing.

## Physical environment

**Water table, and recharge rates**

It can usually be expected that where irrigated agriculture is practised, which meets the conditions listed above, the water table is likely to be closer to the surface than five to six metres. Nevertheless, this should always be verified, and may not be the case where there is the widespread use of a local irrigation technology that delivers water over long distances by canal, but such instances are uncommon in Africa. It is possible, however, that wells may have very low recharge rates, especially if they are dug to provide for very localized irrigation, and required to deliver less than 500 litres per hour. In such a case, dissemination programmes will have to consider what is involved in changing this situation, either by deepening or widening selected wells, or introducing a new technology (such as tubewells). Water table levels and recharge rates should only be verified at the end of the dry season, and before the onset of the rains.

## Economic environment

### Current state and nature of the irrigated agriculture sector

There needs to be a critical mass of farmers already engaged in irrigated agriculture; for sale in local markets for a treadle pump production and marketing programme to succeed. This is because:

- There needs to be an adequate *potential* market for the treadle pump for the programme costs associated with commercialization to be justified. More important, local producers have to be convinced that the investment costs and risks of making the pump are justified by the likelihood of significant sales. For this to take place there needs to be a large number of farmers making a good average income from irrigated agriculture working in a close and accessible geographical zone. If all other conditions are favourable in a given area of concentrated irrigated agriculture, between 10 and 15 per cent of potential users might be expected to have bought pumps after three years of dissemination effort. Planners should determine if this number of pumps is likely to be sufficient to interest potential manufacturers.

- Farmers will not buy a treadle pump just because it is technically superior to alternatives, and saves labour: it has to make sense in terms of increasing *income*, and not merely output. If a farmer does not believe that he or she can make a lot more money using the pump, it will not be bought, despite its many other advantages. Efforts by CARE to popularize the pump in more remote parts of Mali were halted for just this reason.[5] Therefore, the existence of market-based production for cash is an essential prerequisite.

Both of these conditions are generally fulfilled in Mali (12 000 producers), Senegal (60 000 producers), Togo (4 500 producers) North Cameroon (8 000 producers).

### Cost and availability of irrigable land

A programme of treadle pump dissemination is largely justified by its contribution to:

- increased agricultural production
- increased household income.

This can only be achieved if the factors of production are in place, which are:

- land
- labour
- capital.

If there is no possibility of expanding irrigable land, treadle pumps should only be introduced if there is a certainty of greatly increased incomes accruing to farm owners, and a reduced demand for unpaid family labour. If, however, increased production takes place at the cost of reducing the number of *paid* jobs available in the sector, programme planners should carefully reflect upon the likely negative social impact. As a rule of thumb, the technology should only be introduced in places where land availability and cost is not a serious constraint, in order that land under cultivation can increase, and labour saved by the use of the pump can either be applied to increasing the acreage under cultivation, or to alternative economic activity.

ATI's project in Senegal meets both of these criteria. With 850 pumps sold to date[6] treadle pump users have saved an average of 27 person-days of paid labour, and 102 days of unpaid labour per year[7], which would otherwise be devoted to water lifting. There has also been an increase in the mean average size of their gardens from 0.33 ha to 0.46 ha ( a 39 per cent increase). Since yields have also increased, the average increase in farmer income is

[5] Farmers in Macina (Mali) who traditionally practise market gardening have failed to adopt the pump, despite its many advantages, because the sector is extremely competitive, and profits extremely small. Since there is also no cost to family labour (which is available in abundance) they prefer to stick to traditional methods of water lifting.

[6] November 1994

[7] In the case of both Togo and Senegal there has been little apparent labour displacement. In Senegal there has been an average saving of labour required to lift water of 60 per cent. When combined with an average increase of acreage under cultivation of 40 per cent, most of this excess labour has been redeployed to other tasks on the farm. In Togo, where treadle pumps have generally replaced motor-driven pumps, there has been an increase in labour demand.

estimated at $590 (post devaluation). There is an evident contribution to significant rural capital formation,[8] while improving the quality of family life through reduced demand for family labour.

**Cost and availability of labour**

It will, occasionally, be the case that labour is very expensive, especially where there are competing seasonal demands. While this is not often the case, it needs to be remembered that most small-scale irrigated agriculture in Africa takes place close to cities where there may be alternative sources of income, and a higher cost to labour than in rural areas. When this cost is greater than, for example, the cost of fuel, maintenance and/or rental of motor driven pumps, the treadle pump is unlikely to be widely adopted, even where irrigated agriculture is well established.

**Cost and availability of capital**

Despite its low cost compared to motorized pumps, the treadle pump is expensive.[9] While greatly expanding the number of farmers who can now have access to an efficient and cost-effective water-lifting technology, many who can benefit will be unable to afford the pump, since it still represents a significant investment, and credit is often not available to smallholder farmers in sufficient quantities, or at reasonable cost. Programmes *which have carefully analysed the cost-structure of market gardening, and are certain of the existence of positive market conditions* may have to consider various inducements to enable farmers to purchase the pump, and may also have to consider assisting manufacturers to supply pumps on credit.

This can take varying forms. For example, in Senegal ATI initially launched the pump on a cash sale basis. When it became apparent that sales were likely to be slow, the pump was offered with a 50 per cent down payment, and a 50 per cent payment two to three months later (to coincide with harvest). Farmers were told that if after one month they didn't like the pump, they could get their down payment back , but that if they wanted to keep it, they would have to make one or two instalment payments to cover the balance. This strategy was designed to reduce the apparent investment risk associated with an unproven technology, but proved effective as a short-term form of credit, tied to the agricultural cycle. A similar system, managed by agents who were themselves market gardeners, was tried also in Togo, leading to an immediate doubling in sales.

**Profitability of the sector, and capital formation**

When detailed cost and market information are not available, programme planners should look at the evidence of pre-existing investment in the sector. While the treadle pump may enable the establishment of market gardening enterprises where they did not exist before, this is unlikely in the early stages of pump promotion and sales. It is important to focus on a pre-existing sector, and to be reasonably satisfied that not only is it well established in terms of production systems and active markets, but is already capable of significant investment in productive assets.

This involves counting and approximately valuing fixed and current assets, as well as estimating the size of the wage labour force. It is useful also if planners can understand how credit is given and received amongst producers, suppliers, wholesalers and retailers, and whether or not rental of land and equipment is widespread. This information will enable planners to draw up pro-forma cost-structures, and a simplified baseline balance sheet. While obviously imprecise, this data should enable planners to know if the sector is attracting significant investment, whether or not it is profitable, and who those are who most benefit.

If there is clear evidence that the sector is profitable, and has the potential to contribute to the formation of capital, not only is the treadle pump an appropriate product which can increase general levels of profitability and production, but it is likely to find easy acceptance in the market.

**Input availability**

The availability and cost of fuel, spares and agricultural inputs has a profound effect on the ability of programmes to introduce the treadle pump. In Togo, the very high cost of imported spare parts, and the high cost of fuel made market gardeners keenly aware of the advantages of the pump. In

[8] Estimated at $3 500 000, assuming a six-year pump life. *Impact of the Treadle Pump on Market Gardeners in Senegal* ATI August 1994

[9] Before devaluation, when the CFA fr was worth 250 to the US Dollar, a treadle pump in Togo sold for CFA fr 35 000, and in Senegal for CFA fr 38 000. Current price in Senegal is CFA fr 47 000, or $86.

contrast, it has been hard to sell pumps in Northern Cameroon, Nigeria and Niger owing to the very low cost of imported Nigerian fuel, and the very low cost of motor driven pumps.[10]

When inputs are hard to obtain, this may be because there is a generally low level of market-based economic activity, which is the first thing programmers should be alert to. While it may be feasible for projects to procure and supply the required inputs, this must be easily and affordably managed by pump users when programme activities come to an end. Where this is unlikely to happen, pump dissemination should not be attempted.

Where the major constraint to irrigated agricultural production is the supply or cost of agricultural inputs, introducing the treadle pump will likely have only a marginal influence on production. Programmers should discover to what extent such inputs are essential to production, and what is their cost and availability.[11]

## Technical environment

### Raw materials

Raw materials must be easily available to pump manufacturers, so that promotional activities undertaken by projects do not have significant procurement and supply components, except for initial tooling. CARE was successful in introducing a PVC hand pump developed in Bangladesh into the Kanem region of Chad. When CARE's project came to an end producers were unable to obtain the PVC parts except at great cost, and pump production ceased. Even those farmers who had pumps no longer used them for agricultural production, switching instead back to the *Shadoof*, because they feared depending on a pump which could not be replaced or easily repaired. People will not buy a pump if they have reason to believe that they can't either maintain it themselves, or find local services that can do so quickly and at low cost. The establishment of these local services depends on the *pre-existing* local capacity to supply basic raw materials for production.

### Production equipment

It is more important to have local maintenance services than a fully fledged pump producer close by. This is because when a pump breaks down, if it can't be fixed very quickly, the farmer will risk losing an entire season's production. It is therefore important for programme planners to be sure that requisite technical capacity exists in the target zones to repair the pump on a highly localized basis.

Pump producers can be further away from users than maintenance technicians, but do not have to have sophisticated machinery and equipment. The tooling provided by ATI is sufficient to ensure the accurate and rapid assembly of the pump, but pre-supposes that producers have a basic inventory of other equipment. This comprises, *at a minimum* the following:

- Engineer's hammer
- Cold chisel
- Hacksaw
- Flat files
- Hand-operated bench or guillotine shears capable of cutting up to three mm sheet
- Hand-held electric drill
- Pillar drill equipped with vice (a hand-held electric drill can be used, but makes preparation of the valve box extremely difficult)
- Welding transformer (150 amps)
- Bench vice (optional with ATI kit)[12]
- Steel tape measure

In addition, it would be useful (although not vital) to have *access* to an engine lathe in order to prepare the piston disks.

## 2. Promotional activities

Prior to conducting demonstrations it is advisable to answer many of the questions listed above. By

---

[10] The strength of the CFA fr *vis-à-vis* the Naira before the devaluation of early 1994 has made purchase of commodities from Nigeria artificially cheap. The costs of labour and domestically supplied inputs in countries on Nigeria's borders have made pump production and sale problematic. With the devaluation of the CFA franc, this situation may change.

[11] In Togo, CARE promoted the use of the pump in Atakpamé, which is 150 km from the capital city which is the primary market for vegetables. Since Atakpamé is a high altitude region there is a lower need for both fertilisers and insecticides than on the sandy soils of the coast where most irrigated agriculture takes place. The reduced cost of inputs compensated for the higher costs of transporting the harvest to market.

making site visits to irrigated farms it will be possible to learn something about existing practice, and introduction of the pump will be easiest where there is a certain coexistence of manual and mechanised water lifting: that is, where the treadle pump is an intermediate technology. Attempt to identify these cases in advance, so that demonstrations can be conducted where the pump is most likely to be adopted, and where pumping may actually do the farmer some good.

Demonstration of a new technology is the best way of passing along the message of its performance, and suitability in a given setting. A demonstration also makes it possible to target that part of the population most likely to be interested in the technology. For example, a demonstration of the treadle pump set up in an area where the water table is no deeper than seven metres and where irrigated agriculture is already the main source of income is likely to draw a large number of prospective buyers. The demonstration allows the clients to see the pump close up and to evaluate whether or not it responds to their needs. It is also a good idea to leave the demonstration pump in place for a week or two to allow potential buyers the chance to evaluate it for themselves. This avoids the need for repeat demonstrations, and provides time, at the farmer's leisure, for serious thought about pump purchase. Initial reactions from farmers are usually positive, but they do not always reflect their purchasing behaviour. Leaving the pump in place allows time for more solid opinions to form.

Installation of the pump on a well should make use of the type of suction and delivery pipe most likely to be used in the local area, and of the lowest possible cost. It is tempting to use a piece of reinforced flexible hose since it can easily be adjusted to the depth of the well, but it is expensive, and unlikely to be used in practice. More likely is the use of 50mm diameter low pressure PVC pipe, and elbows, joined together with old inner tube. This is low cost and very effective, and a trained field team can set up and adjust appropriate lengths of piping in a matter of minutes.

In choosing a site for the demonstration, it is useful to do so at the home of a dynamic and progressive farmer. Not only does the farmer feel an interest to help in the promotional process, but he or she may have a continuing interest in acting as a sales agent. In both Senegal and Togo these agency arrangements have been vital in creating linkages between pump-makers and clients who live at some distance from each other. This sharply reduces the number of visits that the pump-maker needs to make to develop the market, and is self-financed by payment of small commissions. In Togo, any agent who sells a pump receives a CFA fr 3 500 commission, which is added on to the price of CFA fr 35 000. In Senegal the commission is currently CFA fr 5 000 over the average wholesale price of CFA fr 42 000.

Finally, to ensure the best possible reception of the technology, it is important to conduct demonstrations with the permission of, and preferably with the participation of, local government officials and traditional authorities.

---

[12] The tooling illustrated in this manual shows a vice which is provided with the toolkit. Because this vice is quite expensive, present versions of the toolkit do not now include this vice. Instead, the tooling is made with plates welded vertically to the underside, permitting the tools to be clamped into any standard heavy-duty bench vice. Since nearly all metalworkers possess such vices, this reduces the overall cost of the tools.

# Cost Issues

## 1. Cost comparisons

### Pump prices

The cost of producing a treadle pump mainly depends upon the local cost of its raw materials and of the labour required in its manufacture, as well as the efficiency and scale of production. In Mali, Cameroon and Togo the initial retail (ex-factory) price set by producers was CFA fr 32 000, CFA fr 35 000 and CFA fr 35 000 respectively.[13] By comparison, pumps made in Bangladesh have sold for about US$20. This enormous difference is partly accounted for by differences in design between the two pumps. The present model of the treadle pump is, as already noted, produced complete with treadles, support frame, and double-acting piston/cylinder set, whereas the Bangladesh pump consists only of a single acting piston/cylinder set. More important, however, there is a great difference in cost factors. Furthermore, the large market that the pump has enjoyed in North Bangladesh, and its production by a large number of artisans, has likely caused an almost perfect competition amongst producers. This effect could lead to reduced prices also in Africa, but not nearly to the same extent as in Bangladesh, where the density of population and absolute demand for the pump is much higher. In addition, raw materials are cheaper in Bangladesh than in Africa. In Togo, where materials are cheaper than in surrounding countries to the north, raw material costs were estimated at CFA fr 16 000 in 1991 (roughly $64).

*How does the cost of the treadle pump compare with that of the rope and bucket and the motorized pump?*

We have examined this question for both Mali and Northern Cameroon[14]. Sprinkler cans are produced in Mali by local artisans from galvanized sheet steel, and cost between CFA fr 5 000 and CFA fr 8 000. Since the rope used is short, its cost can be taken as negligible. Motorized pumps are available from two distinct sources: the few large stores in Bamako that import goods through official channels, and the relatively large number of smaller shops and itinerant traders that sell goods smuggled into Mali from Nigeria or The Gambia. Pumps sold by the large stores cost at least CFA fr 300 000 ($475), this being the price of a gasoline (petrol) driven 2" (50mm) centrifugal pump that has a rated delivery of 17 m$^3$ per hour at a lift of five metres. The suction hose, foot valve and 50 metres of delivery hose for this pump cost an additional CFA fr 150 000, raising the total cost to CFA fr 450 000 ($785). Larger pumps are proportionately more expensive: a 3" (75mm) gasoline pump costs CFA fr 620 000 with its suction hose and foot valve, but no delivery pipe, and a 4" (100mm) diesel pump costs CFA fr 2 800 000 with no suction hose, foot valve or delivery pipe. The smaller shops are somewhat less expensive, selling the smallest gasoline powered pump at CFA fr 200 000 with no suction hose, foot valve or delivery pipe. In North Cameroon, where black market and official suppliers also coexist there is a much larger difference between their prices. This is because the government of Cameroon imposes much higher import duties (up to more than 100 per cent), and because the North Province of Cameroon shares a long border with northern Nigeria. Goods that are imported to or produced in Nigeria are sold in Cameroon at a premium to obtain hard currency (CFA franc). The lowest-cost motorized pumps (2" gasoline, centrifugal) can be purchased on the black market for CFA fr 150 000, with no hose or foot valve, and in good used condition for even less. The same pump sold in stores costs CFA fr 373 000. As already discussed (under Input availability page 8) we believe this can seriously threaten the viability of treadle pump production in North Cameroon. ATI's prospective producer of pumps in North Cameroon faces the same problem when selling his bicycles. Bicycles imported on the black market sell for less than the cost of his raw materials.

These kinds of price distortions are common in Africa and need to be understood and carefully measured when assessing the viability of treadle pump production. The fact that the formal and informal sectors can coexist when their prices are

---

[13] Roughly $100–110 at the time of production start-up prior to devaluation. Although prices have increased up to CFA fr 47 000 since then, the dollar denominated cost has dropped to about $85.

[14] 1992 pre-devaluation prices.

so very different shows that they cater to different markets. The formal sector supplies government, parastatals and international development agencies, plus a large part of the formal private sector. Although the eventual market for treadle pumps made in North Cameroon will be constrained by competition from motorized pumps imported on the black market, the immediate market is, in part, those institutions that are obliged to purchase from the formal commercial sector.

When these costs are compared, the treadle pump, which sells at about CFA fr 47 000 is about 30 per cent of the cost of the lowest cost gasoline-powered alternative: a 2" motorized pump with a capacity of 8 cubic metres per hour costing CFA fr 150 000 on the black market. The same pump costs about CFA fr 373 000 in other parts of West Africa under normal conditions of sale, and is therefore nearly eight times more expensive than the treadle pump.

We have not compared the treadle pump with other human-powered pumps that are already on the market in Cameroon, Mali, Senegal and Togo because, as we have already shown, they are comparatively expensive, and ill-suited to irrigation. No hand pump or foot-powered pump is currently in widespread use for irrigation in any of these countries.

**Viability**

In addition to considering the initial or capital cost of the alternative water lifting methods, it is important to consider their operating costs, and thus to determine the viability of competing systems.

A detailed financial analysis of the treadle pump versus manual lifting of water from a well by a bucket and sprinkler can, and a small motorized pump has been conducted for market garden sizes ranging from 0.10 to 0.67 ha in Senegal. This study found that the treadle pump was the most financially beneficial of the technologies under various scenarios of wage rates, land costs, financing and well and basin costs at real discount rates of 10 per cent and 20 per cent. This analysis was confirmed by before-and-after comparison of the incomes of treadle pump owners.

When the treadle pump is introduced in a farm which has traditionally lifted water by rope and bucket the biggest impact is felt in terms of reduced costs for labour. Based on our interviews in Mali, Togo, and Senegal it appears that seasonal farm labour costs between CFA fr 500 and CFA fr 800 per day. Roughly half of the labour used in market gardens is for irrigation. Assuming that the treadle pump can cut this labour requirement approximately in half,[15] it would appear that the pump could save from CFA fr 125 to CFA fr 200 per day per labourer employed, or from CFA fr 250 to CFA fr 400 per day, per average farm of 0.15 hectares. In a four- month season, it could pay for itself in the first year.

If these labour savings are applied to increasing the acreage under cultivation, vastly greater benefits are likely to be felt. Annual income increases measured before and after adoption of the pump in Senegal have averaged CFA fr 323 492 ($650).

*How viable is the treadle pump compared to the motorized pump?*

A quick comparison of the costs of lifting water for irrigation with the treadle pump and a small motorized pump is shown below for market gardens of 0.25, 0.50 and 1.0 hectares. The 0.50 hectare garden requires 25.0 cubic metres of water per day. It is assumed that irrigation is needed for 180 days per year, resulting in a total water requirement of 4 500 cubic metres per year for 0.50 hectares. The lifted water is distributed under pressure through PVC pipe. The cost data are from Senegal in mid 1994 (post devaluation).

**Treadle Pump**

- The treadle pump is worked by two people at the same time and delivers six cubic metres of water per hour of operation. Thus 4.2 hours of labour are needed for two persons at CFA fr 100 per person-hour. The total labour cost is CFA fr 840, which is CFA fr 33.6 per cubic metre of water.

- The treadle pump costs CFA fr 47 000, and an additional CFA fr 25 000 is needed for 50 metres of PVC pipe. Since the pump and pipe have an expected lifetime of six years, the annual depreciation would be CFA fr 12 000. The annual depreciation is equivalent to CFA fr 2.7 per cubic metre of water lifted.

[15] See footnote 7

- Spare parts and repairs cost CFA fr 3,500 and lubricant CFA fr 13,500 per year. The total maintenance and repair cost is equal to CFA fr 3.8 per cubic metre of water lifted.

- The total cost of water for 0.5 ha with the treadle pump is, therefore, CFA fr 40.1 per cubic metre.

**Motorized Pump**

- The small motorized pump, with an outlet pipe diameter of 40mm, costs CFA fr 373 000 and has an expected lifetime of four years. The annual depreciation for the pump is, therefore, CFA fr 93,250.[16] An additional CFA fr 100 000 is needed for the suction hose, foot valve, and 50 metres of delivery pipe, which are depreciated over six years, resulting in an extra annual depreciation of CFA fr 16 700. The total annual depreciation of CFA fr 109 950 amounts to CFA fr 24.4 per cubic metre.

- Spare parts and repairs cost CFA fr 60 000 per year, or CFA fr 13.3 per cubic metre of water.

- Since the capacity of this pump is eight cubic metres per hour, it would be operated by one person for 3.1 hours per day, at a labour cost of CFA fr 310 per day. The labour cost is CFA fr 12.4 per cubic metre of water lifted.

- The fuel cost is 0.5 litres of gasoline (petrol) per hour for 3.1 hours at CFA fr 455/litre. The fuel cost per day is CFA fr 705 per day, or CFA fr 28.2 per cubic metre of water.

- The lubricating oil requirement is one litre per 20 hours of operation, at a cost of CFA fr 750/litre. The daily cost of lubricating oil is CFA fr 116, which is CFA fr 4.6 per cubic metre of water.

- The total cost of water for irrigating 0.5 ha with the motorized pump is, therefore, CFA fr 82.9 per cubic metre.

Table 1 below compares the cost per cubic metre of water delivered by the treadle pump with water delivered by a motor-driven pump under three different conditions: irrigation of 0.25, 0.5 and 1.0 ha respectively. It shows that while the comparative operational cost advantage of the treadle pump diminishes slightly as the hectarage increases, it still costs only about half as much to lift water using the treadle pump compared to the motor-driven pump.

These assumptions can be modified somewhat, giving slightly higher costs to either the treadle pump, or the motorized pump, but we conclude that for the conditions prevailing in Senegal the cost per cubic metre of water delivered are clearly lower for the treadle pump. Very similar figures pertain in Togo and elsewhere in coastal West Africa.

[16] Depreciation is calculated as a combined function of pump use and pump age. It is not, therefore a linear function of total water pumped.

**Table 1: Cost comparison per cubic metre of water pumped**

| | 0.25 ha | | 0.50 ha | | 1.0 ha | |
|---|---|---|---|---|---|---|
| **Cost per cubic Metre of Water Pumped** | **Treadle Pump** | **Motorized Pump** | **Treadle Pump** | **Motorized Pump** | **Treadle Pump (2)** | **Motorized Pump** |
| Labour | 33.6 | 12.4 | 33.6 | 12.4 | 33.6 | 12.4 |
| Depreciation: | | | | | | |
| 3 Years | | | | | | 15.7 |
| 4 Years | | | | 24.4 | | |
| 6 Years | | 35.0 | 2.7 | | 2.7 | |
| 8 Years | 4.0 | | | | | |
| Spares & Maintenance | 3.8 | 13.3 | 3.8 | 13.3 | 3.8 | 13.3 |
| Fuel | | 28.2 | | 28.2 | | 28.2 |
| Lubricants | | 4.6 | | 4.6 | | 4.6 |
| **Total** | **41.4** | **93.5** | **40.1** | **82.9** | **40.1** | **74.2** |

## Table 1: Cost comparison per cubic metre of water pumped

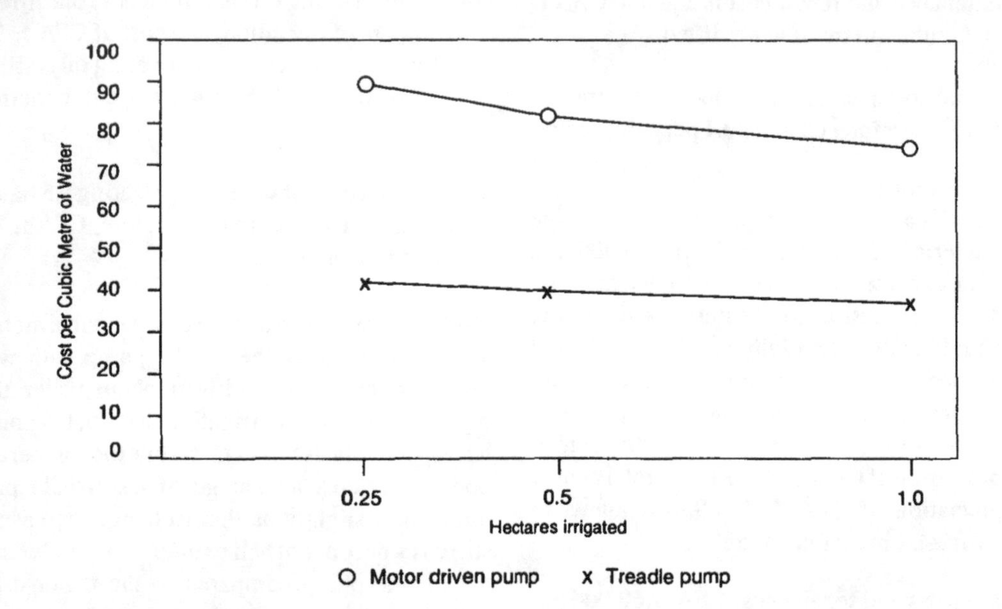


In Northern Cameroon, with much cheaper costs to the motorized pump (CFA fr 150 000), there would be no discernible advantage to either technology. Note however, that while the lion's share of the treadle pump cost is for labour, and very little for the cost of materials, nearly the entire cost of the motorized pump is for imported commodities: either capital cost, or fuel and oil. There is, therefore, a clear economic advantage in favour of the treadle pump.

This sort of cost comparison is important to make before initiating a programme of pump manufacture and promotion, because the potential buyer of the pump has to balance many different factors in making a purchase decision, only one of which will be the initial cost of the pump. When both purchase price and operational cost per cubic metre are both clearly in favour of the treadle pump (which is true in nearly all the cases studied) prospective owners will likely then be influenced against the pump only if it is awkward to use, proves unreliable or difficult to repair, or makes organizational demands *vis-à-vis* a large labour force that he or she would rather not confront.

The sort of data comparison shown here should be prepared as part of the information materials used in marketing.

# Treadle Pump manufacturing instructions

## General

This section of the manual is designed for the manufacturer of the treadle pump, or for training institutions that want to help small workshops start making the pump. It is assumed that prospective manufacturers will make use of the tooling provided by ATI, around which the text and illustrations are based. Without this tooling it will be virtually impossible to manufacture the valve-box, which requires precision in the distribution of the valve port honeycomb, and hard to guarantee the regularity of the pump cylinders.

The text and illustrations are organized into the following sections:

- *Making the pump cylinders*
- *Making the pump body*
- *Making the handle and treadle support*
- *Making the pistons, piston rods and pump cups*
- *Making and installing the valves*
- *Making the wooden parts of the pump*
- *Final assembly of the pump*

Wherever possible, the illustrations have been matched to the text, and organized into groupings (*Figures*) which show a complete process, through a numbered series of drawings captured in a single frame. For the sake of clarity, most drawings do not show dimensions. Wherever necessary, these are to be found in the text, together with material specifications, types of welding rod and amperages required. These have been italicized in bold face print, to enable them to be easily identified. Bold face italics are also used to emphasize points of particular importance.

# 1. Making the pump cylinders

The cylinders of the treadle pump are made from mild steel sheet which is rolled around a steel pipe mandrel *(Figure 1, Drawing 2)*. The seam is subsequently arc-welded, while the mandrel remains inside the cylinder. This method produces cylinders that, while not perfectly round, are smoother, lighter, and far less expensive than cylinders made from commercial steel pipe. The flexibility of the leather pump cups enables them to effect a good seal, even if the cylinders are slightly out-of-round.

***Each cylinder is made from a piece of 1.5mm (1/16"), or 2.0mm thick sheet steel, measuring 330mm x 330mm.*** The 1.5mm thickness is best, because the sheet can be easily bent around the mandrel, yet is thick enough to be arc-welded without great difficulty, and without causing distortion. The dimensions of the sheet were selected because they enable 18 cylinders, or nine pumps to be produced from a single sheet measuring one by two metres, while giving the cylinders a suitable diameter and stroke. Because the cylinder sheet (or blank) is square, the pump maker cannot confuse the length and width of the sheet when rolling a cylinder, and thus cannot produce cylinders that are the wrong size. The 1.5mm thick sheet is easily cut on a manually operated bench shear, of the type often used, and sometimes made, by metalworking artisans in developing countries. Care should be taken to measure and cut the cylinder sheets as accurately as possible, making the edges very straight and square, to facilitate the welding of the rolled cylinders. To do this, the cylinder sheets should be measured using a good quality steel rule and accurate straight edge, clearly scribed, and checked for squareness before cutting.

A special cylinder rolling tool *(Figure 1, Drawings 1a, 1b and 2)* is included as part of the pump production toolkit. The tool consists of three parts:

- the bracket that supports the mandrel on the steel vice *(Figure 1, Drawings 1a and 1b)*
- the mandrel, used for bending the cylinder blank *(Figure 1, Drawing 2)*
- the bending roller, which helps in rolling the cylinder sheet, particularly its last 150mm *(Figure 3, Drawing 1a)*

When in use, the bracket is bolted to the steel vice (also provided as part of the toolkit), in place of the removable hardened steel jaw on the movable face of the vice *(Figure 1, Drawings 1a and 1b)*[17]

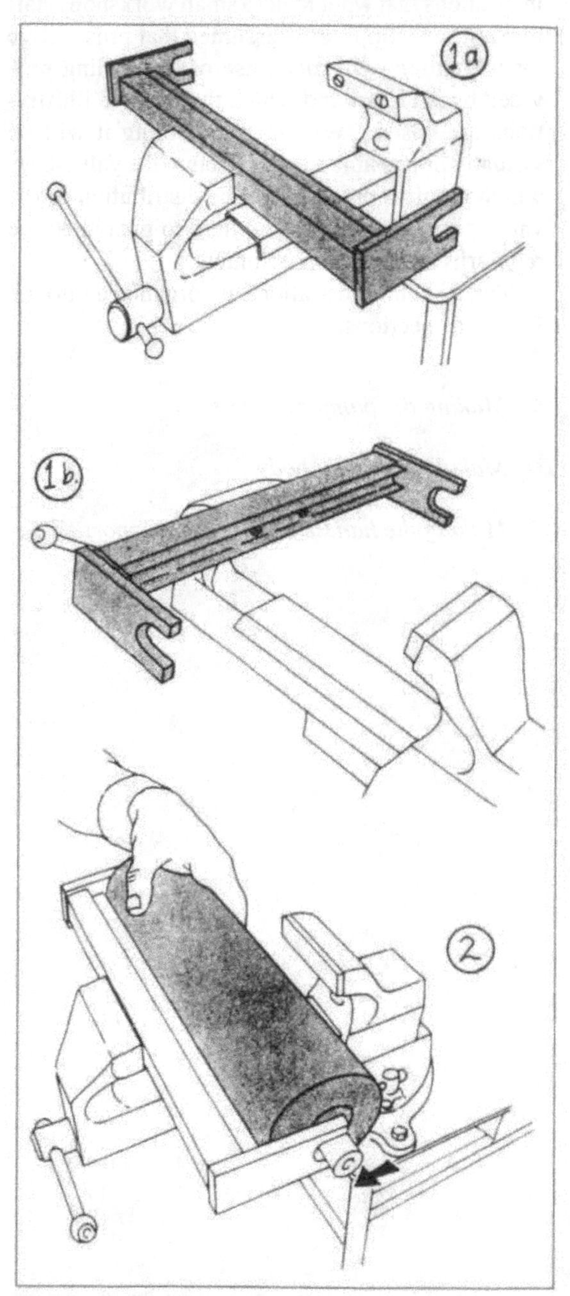


***Figure 1: Cylinder rolling tool kit***

[17] The latest version of the rolling tool is clamped directly into any vice, which is no longer supplied with the toolkit.

The cylinder sheet is inserted between the mandrel and the bracket, so that an edge of the sheet lines up with the lower edge of the bracket *(Figure 2, Drawings 1a and 1b)*. The vice is tightened to clamp the sheet in this position *(Figure 2, Drawing 2)*. Tightening the vice bends the first few centimetres of the cylinder blank, which begins the rolling process.

With the vice securely tightened, push the top edge of the cylinder blank towards the bench, until it touches the body of the vice *(Figure 2, Drawing 3)*. Next, loosen the vice one to two turns, and rotate the blank downward around the mandrel, so that the unrolled portion begins just above the mandrel bracket *(not illustrated)*. Make sure that the blank is properly positioned and remains square

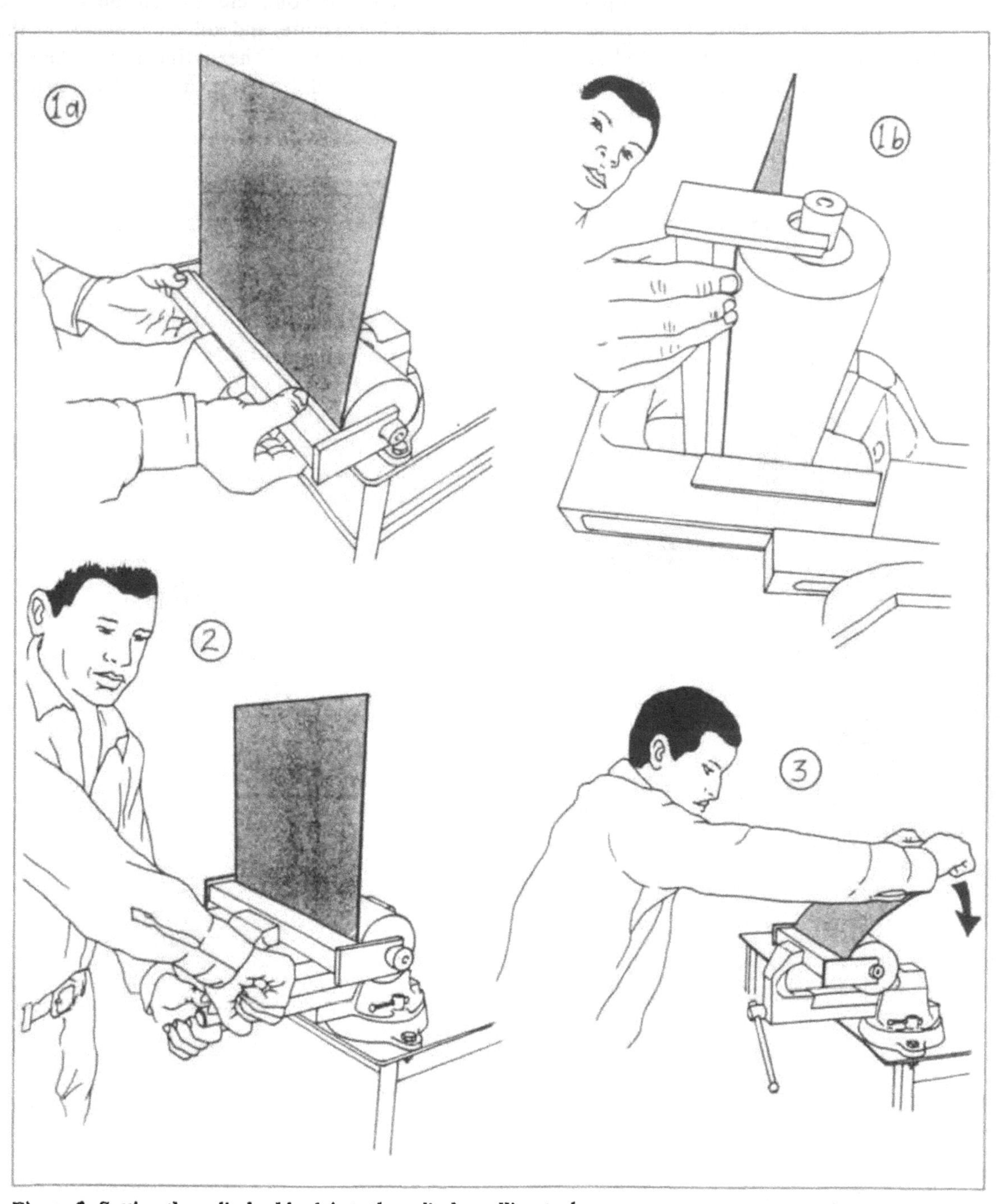


***Figure 2: Setting the cylinder blank into the cylinder rolling tool***

in the mandrel by pulling it upwards, so that its rolled portion contacts the underside of the mandrel along its entire length. As the cylinder develops, it is advisable to tap it upwards into position against the mandrel, prior to tightening of the vice. Tighten the vice again, and once more push the blank forwards. As the cylinder develops, the amount of sheet remaining becomes shorter as it is fed around the mandrel, and at this point it becomes necessary to use the cylinder roller. Fit the roller over the shaft of the mandrel *(Figure 3, Drawings 1b and 1c)*. Push the handle of the roller towards the bench, and verify that the hooks are fully seated on the shaft, using a hammer and a block of wood to tap them downward if necessary *(Figure 3, Drawings 2 and 3)*.

Continue loosening the vice, and feeding the cylinder further around the mandrel, using the roller *(Figure 4, Drawings 1,2, and 3)*. Oil the shafts of the mandrel and the hooks of the roller so that the hooks turn freely on the shaft. When the cylinder is complete, loosen the vice, and remove the mandrel and rolled cylinder from the mandrel support. The cylinder can then be easily removed from the mandrel *(Figure 4, Drawings 4 and 5)*.

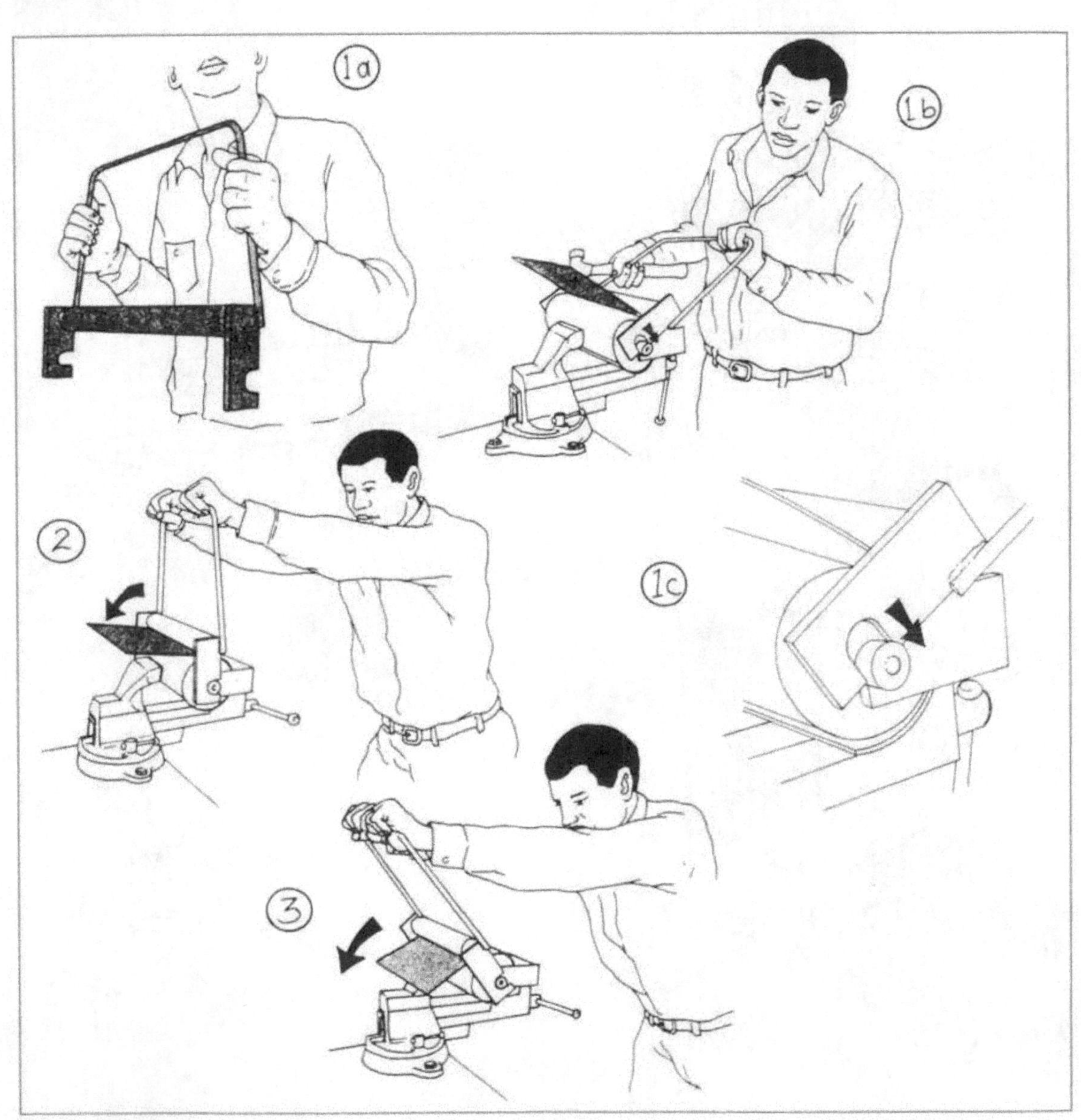


***Figure 3: Using the cylinder roller***

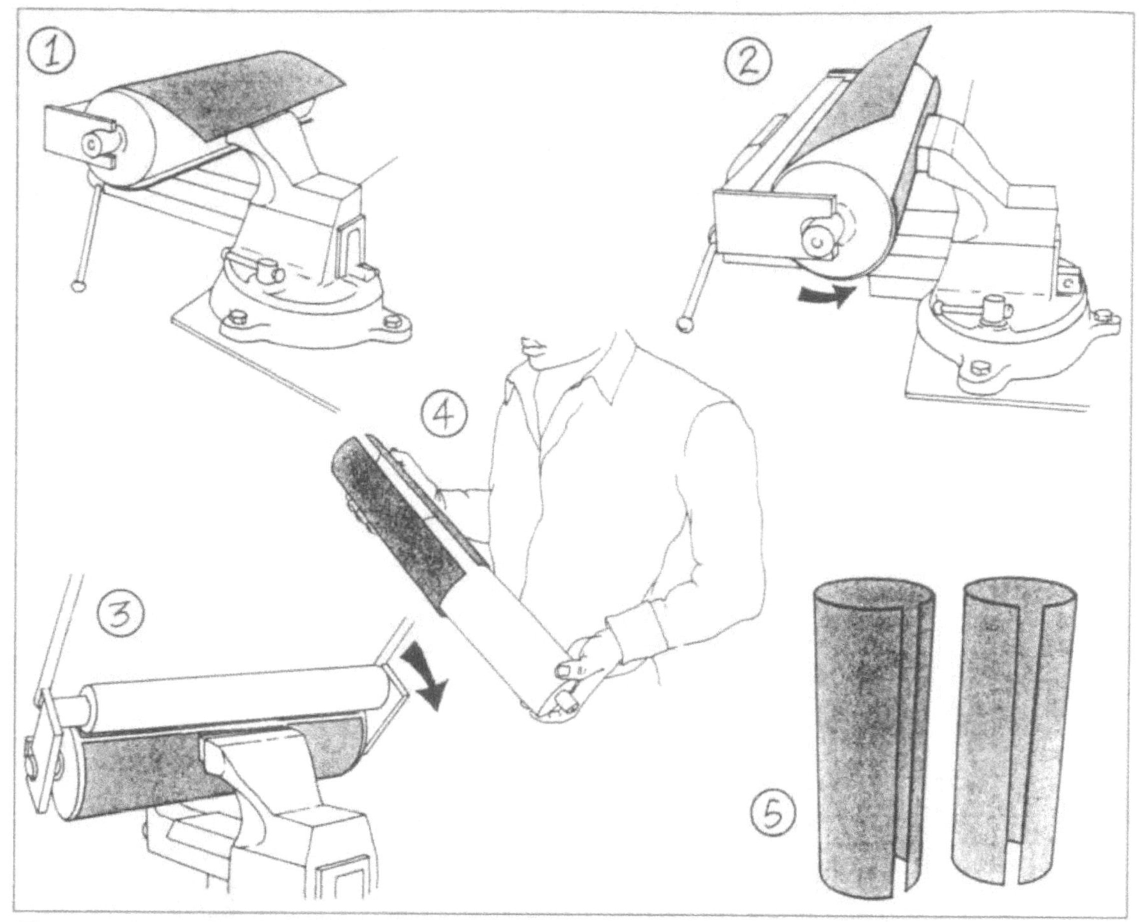


*Figure 4: Completing the rolling of the cylinder and removal from the tool*

Do not be concerned if the cylinder blank is not perfectly round, nor be concerned if its edges do not come exactly together. Since the cylinder seams are welded while mounted on the mandrel, their shape and alignment can be corrected at that time. When making your first cylinders roll two or three, and then proceed to the welding process. This will show how well cut and formed the cylinder blanks need to be. Once one or two cylinders have been successfully welded, a large number of blanks can be cut and rolled, and welded after they have all been rolled. By producing a large number of parts in a batch, the time required to produce each pump can be greatly reduced, at a better quality.

The next step is to tack-weld each cylinder seam. It is best to do the welding on a different vice than the steel vice used with the toolkit, so that the steel vice is maintained in good condition for the work that it can do best. Any vice that opens to at least 125mm (five inches) can be used to do the cylinder welding.

Place the mandrel inside one of the rolled cylinder blanks so that the seam faces upwards. Position the blank between the jaws of the vice, so that the vice is approximately at the middle of the cylinder, and tighten the vice just enough to hold the blank and the mandrel in position. Check that the top and bottom edges of the cylinder are in alignment, and, if they are not, lightly tap them into position *(Figure 5, Drawing 1)*. Now use the special vice grip furnished with the toolkit to clamp the cylinder seam firmly against the mandrel *(Figure 5, Drawing 2)*. The edges of the cylinder seam should butt snugly together, and lie flat on the mandrel on either side of the vice grip. This is vitally important, because if there is a gap along the line of the seam there is a danger that welding along the joint line will attach the cylinder to the mandrel.

Tack-welding the cylinder seam is best done with ***a 2.5mm (3/32") diameter mild steel elec-***

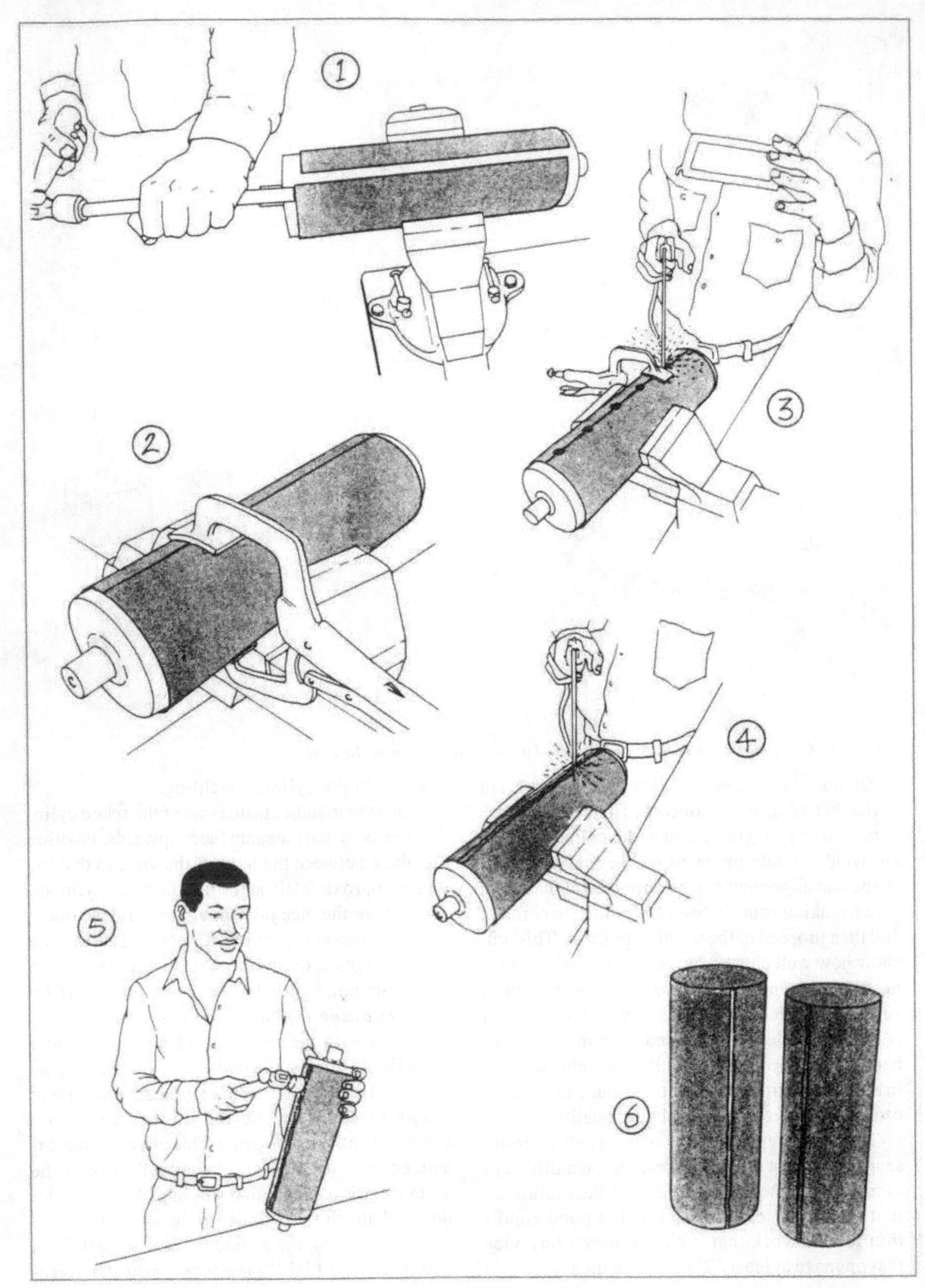


***Figure 5: Welding the cylinder seams***

***trode (type 7014 or equivalent), and the welding transformer set to 75 amps.*** Hold the electrode in a roughly vertical position while tack-welding the seam *(Figure 5, Drawing 3)*. Tack-weld the cylinder on each side of the vice grip before moving the grip to a new location along the seam *(Figure 5, Drawings 2 and 3)*. The seam should be tacked with a total of about eight tacks, spaced approximately 40mm (1 1/2") apart. If, while tack-welding the cylinder seam, the seam bulges away from the mandrel ahead of the welds, or fails to come snugly together, correct such defects before continuing to tack. Usually, slight bulges or gaps at the seam can be corrected by tapping the seam lightly with a hammer, and by tightening or loosening the vice-grip.

The cylinder is now ready for welding. This is most easily done while the cylinder remains on the mandrel. Not only does this provide a secure and stable mounting for the cylinder during welding, but it also prevents the cylinder distorting. This is because the mandrel rapidly conducts heat away from the weld. ***Use the same mild steel electrode as for tack-welding, i.e. of 2.5mm (3/32") diameter, and set the welding transformer to 75 amps.*** It is also useful to tip one end of the cylinder slightly downwards at about 15° to the horizontal, and to weld in the downhill direction towards the welder, but with the electrode facing slightly uphill *(Figure 5, Drawing 4)*. The electrode should be moved along the seam quickly, but steadily, to produce a secure seam, with no burn-through. Using a new tinted glass in the welding mask will make the seam easier to see, and will greatly facilitate cylinder welding. With practice, it should be possible to weld the entire seam with a single electrode, and without stopping.

The cylinder is removed from the mandrel by removing both from the vice, holding the mandrel in a near-vertical position and tapping along the seam with a hammer *(Figure 5, Drawing 5)*. Doing this will flatten out the bulge that occurs along the seam, and will loosen the cylinder from the mandrel, causing it to slide off. If, for whatever reason, the cylinder will not come off, it can be removed by cutting along the length of the cylinder with an electric angle-grinder (but **not** along the seam, since this will be rough, and make control of the angle-grinder more difficult). When the cylinder is split in this way it can be peeled off the mandrel. This should be done with great care to avoid cutting a deep groove in the mandrel. The size of the sheet steel cylinder blank is such that the cylinder should fit somewhat loosely over the mandrel, and it should not normally be difficult to remove. Completed cylinders should be checked to ensure that they stand perpendicular to a flat surface before being set aside in storage. *(Figure 5, Drawing 6)*.

After removing the cylinder from the mandrel, inspect the weld both inside and out. The inside seam should be smooth, with no defects that could tear the leather pump cups, or prevent them from sliding. Any roughness should be filed smooth with a half-round file, followed by polishing with a medium grade of emery cloth. After doing this, check that there are no holes in the weld that could cause leaks. These can be filled by fitting the cylinder back on the mandrel, and welding the hole ***with the transformer set to a slightly higher amperage.*** After filling the hole, remove the cylinder, and inspect it again for internal defects.

With experience, the entire process of cutting, rolling, tacking and welding a cylinder can be done in less than 30 minutes. When pumps are produced in large numbers, and each of these steps are performed on all of the cylinders in batches, the time for producing each cylinder can be reduced to roughly 15 minutes, or about half an hour for the two cylinders required for each pump. If the pumps are made at a large metalworking shop, the use of a guillotine shear for cutting the blanks can further reduce the production time.

# 2. Making the pump body

This section covers all of the steps that are needed to make the pump body, which is the assembled unit that includes:

- o Cylinders
- o Valve-box
- o Inlet and outlet pipes

Whether the pump is produced by a large, well-equipped metalworking shop, or by a small-scale artisan, these instructions are designed to be used with the basic kit of specialized pump-production tools, obtainable through ATI.

The first part of the pump body to make is the valve plate, which is the top of the valve box to which the cylinders are welded. The toolkit includes a pattern which is used for marking out the plate. As with all other parts that are cut from sheet, the blank for this part should be carefully measured, clearly scribed, and checked for squareness. ***The blank is cut from 3mm mild steel sheet, as a rectangle measuring 130mm by 240mm.*** The pattern is then clamped securely to the top of the blank *(Figure 6, Drawing 1)*. The outline of the valve plate is scribed onto the blank *(Figure 6, Drawing 2)*. Use a hand-operated bench shear *(Figure 6, Drawings 3 and 4)* to cut the rectangular blank along the scribed line. The curved corners can be quickly and accurately cut on the hand shear by rotating the blank towards you as the cut is made. If the pump is being made by an artisan who does not own a bench shear, the blank can be cut using a hammer and cold chisel. If this method is used it is important to file the rough edges of the blank after cutting is complete.

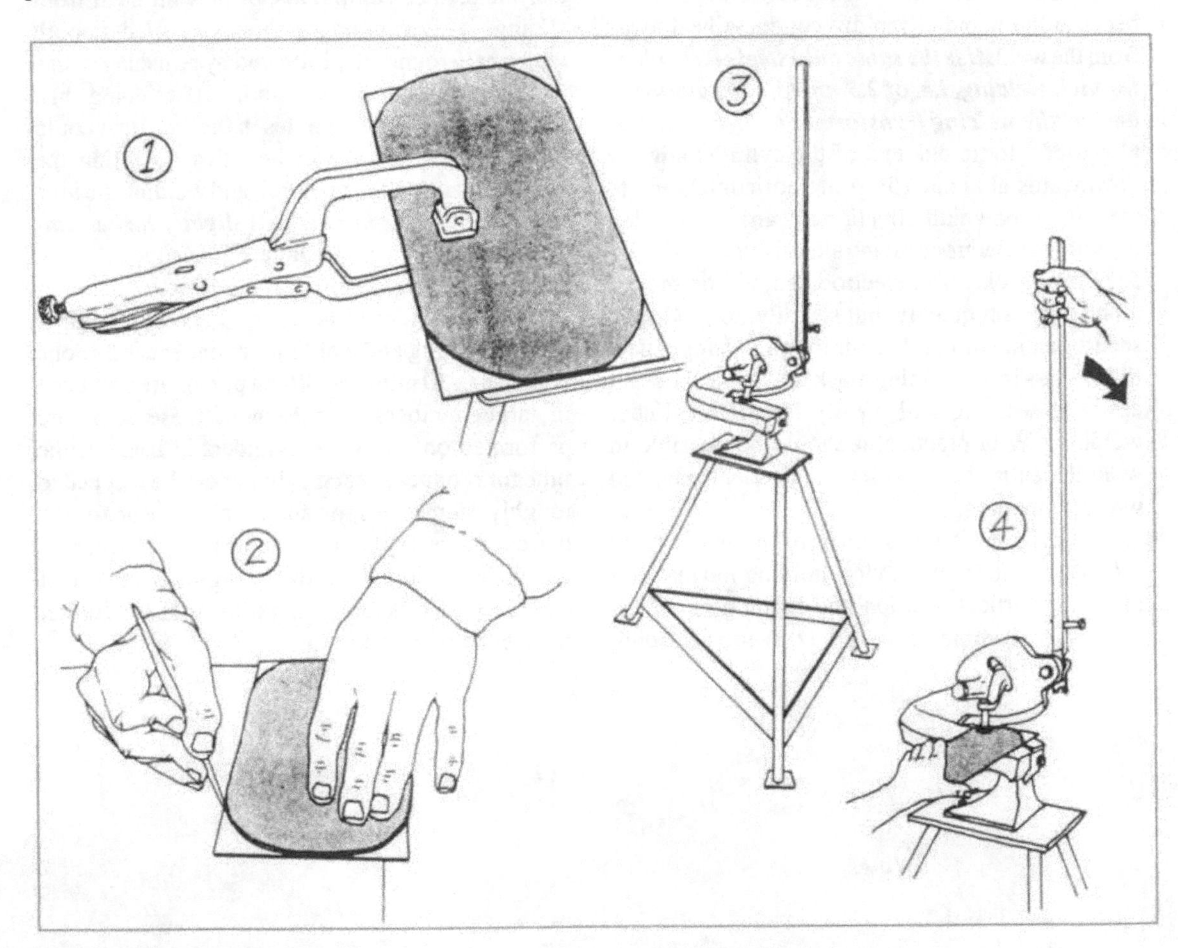


***Figure 6: Cutting the valve plate***

The next step is to mark out and drill into the valve plate the holes that form the valve ports. ***Each valve port is made up of 12 holes of 12mm diameter***. Because it is important to have as many holes as possible, to enable water to flow easily through the valves, it is important that the layout and positioning of these holes in the valve plate is as precise as possible. In order to achieve this, the position of these holes is marked by means of a jig, illustrated below *(Figure 7, Drawings 1a and 1b)*.

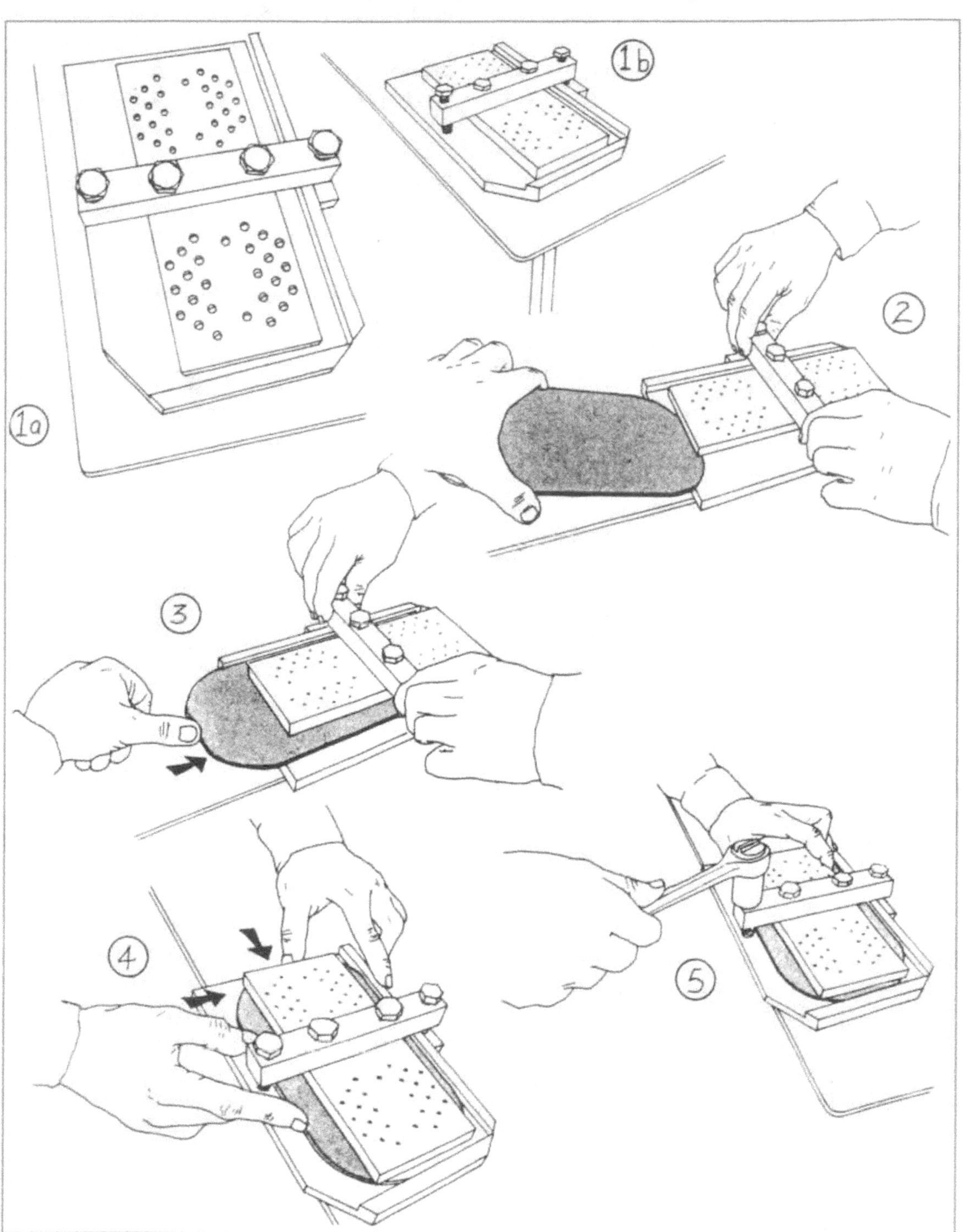


***Figure 7: Valve port marking jig, and setting the valve plate in place***

The two outer bolts on the marking jig are loosened enough to permit the valve plate to be slid between the upper and lower plates of the jig *(Figure 7, Drawing 2)*, from the open end. The plate is slid along the jig *(Figure 7, Drawing 3)* and pushed against the end-wall and side-wall *(Figure 7, Drawing 4)*. When the valve plate is resting snugly in position the two outer bolts are tightened, securely clamping the plate in position *(Figure 7, Drawing 5)*.

The plate is marked by means of a pin punch, which is supplied with the toolkit. A pin punch differs from an ordinary centre-punch. While a centre-punch has a tapered shank, a pin punch is precision ground to a fixed diameter along the length of the shank. A pin punch is used because an ordinary centre punch can be entered into the marking jig guide-holes at an angle. Because a pin punch has parallel sides it can only be entered into the guide-holes in a vertical position, and will therefore enable the holes to be more accurately positioned. A pin punch is illustrated below *(Figure 8, Drawing 1)*. Pin punches are difficult to obtain in developing countries, but an adequate substitute can be made from a 6.5mm drill shank, ground to a point, similar to that illustrated in Figure 8, Drawing No. 1 (inset) below.

The punch is dipped in oil, slid into each of the holes in the jig, and smartly tapped ***once*** with a hammer *(Figure 8, Drawing 2)*. When all of the holes have been marked in this way, the outer bolts of the marking jig are loosened, and the valve plate removed. The plate will carry centre-punched marks, to permit accurate drilling of the valve ports, and the holes which allow the valve rubbers to be clamped to the valve plate.

The valve plate has four sets of marks, which will be drilled to form the inlet and outlet valves in

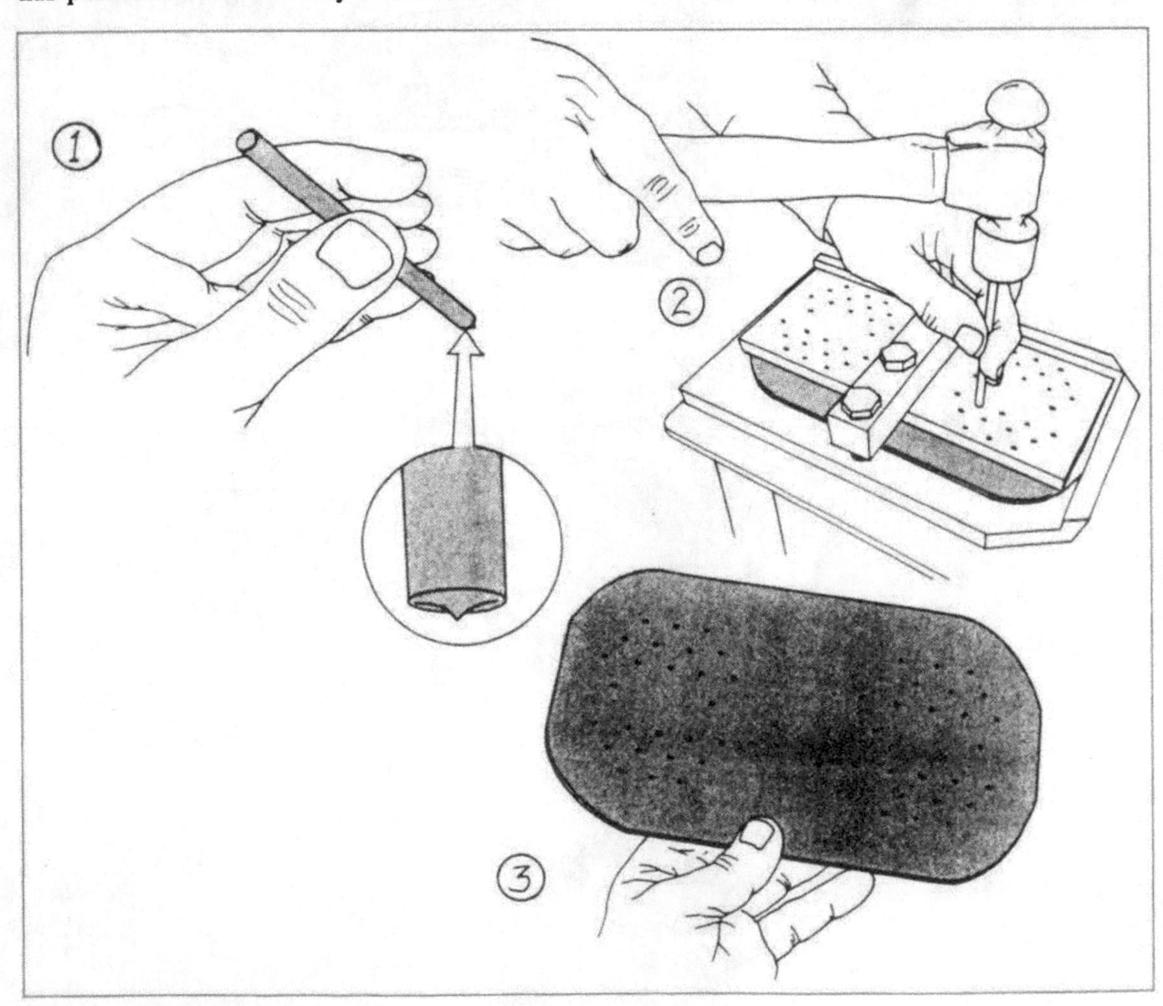


***Figure 8: Punching the valve port hole centre points***

the two cylinders. Between these groups of marks, and along the centre-line of the plate are four other marks. ***These are drilled with a 6.5mm drill***, and the outer holes used to clamp the plate to a flat block of wood by means of wood screws *(Figure 9, Drawing 1)*. The block of wood is then clamped into a machine vice, mounted loosely on the table of a pillar drill.

***Under no circumstances should the operator attempt to drill the valve plate without attaching it to a block of wood: holding the plate by hand can lead to serious hand injury, and equipment damage.***

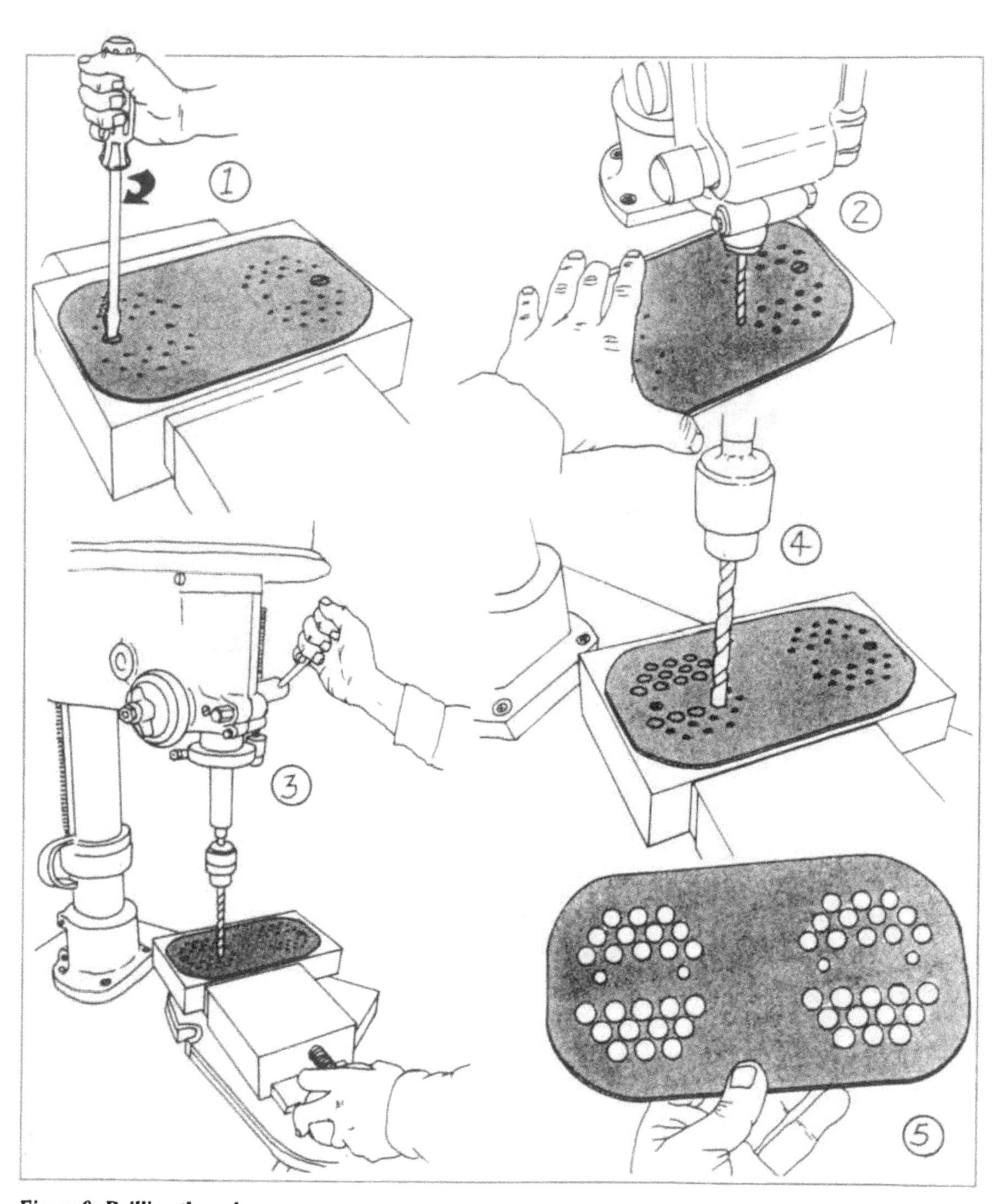


***Figure 9: Drilling the valve ports***

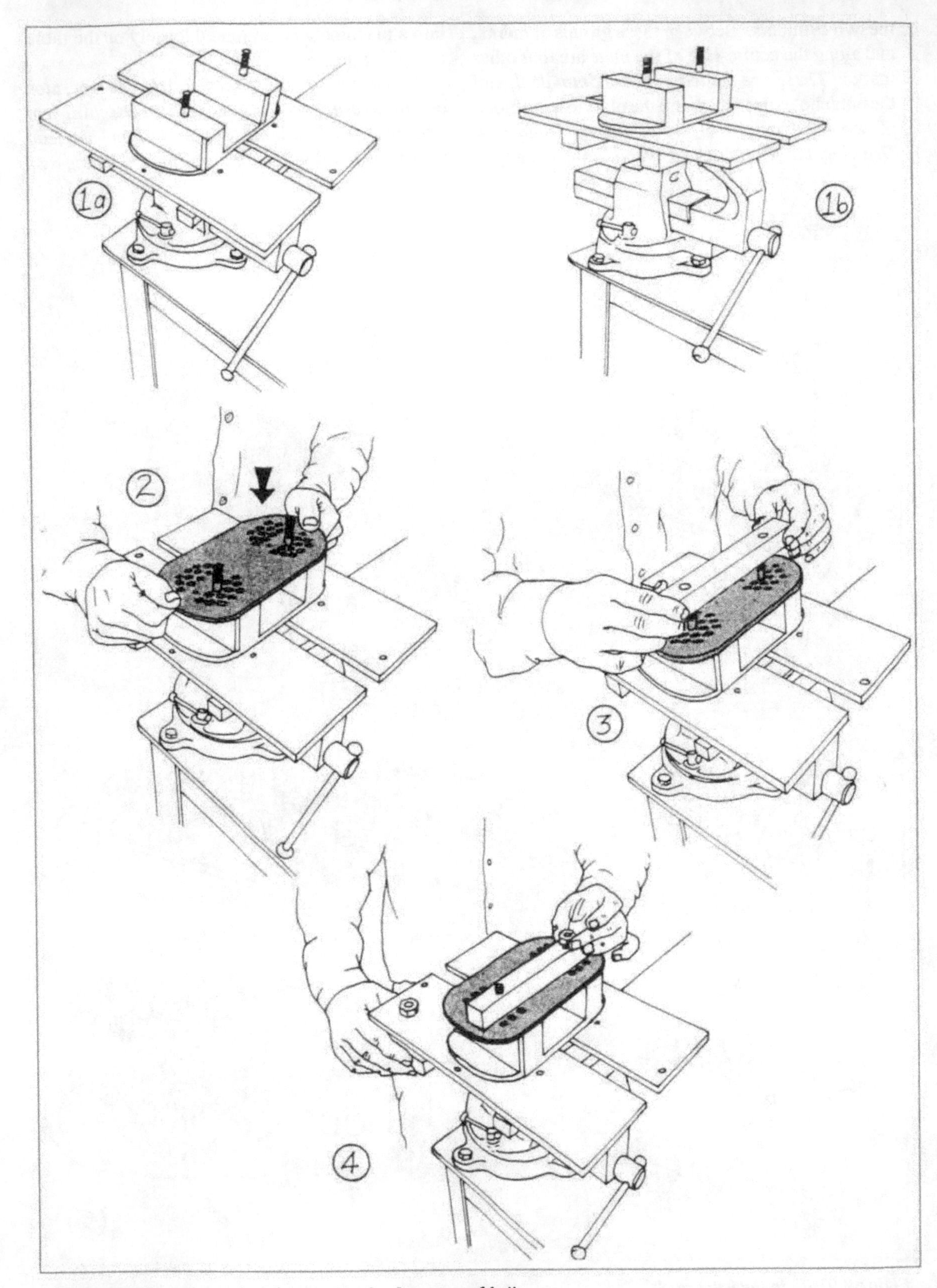


***Figure 10: Mounting valve plate onto valve box assembly jig***

The valve ports are not drilled immediately with a full-size bit *(12mm)*. If this is done the bit will tend to chatter, leaving an irregular, oversize hole. To avoid this, ***the ports are first of all drilled with a 6.5mm drill bit, at a speed no greater than 600 rpm, and frequently lubricated with motor oil*** *(Figure 9, Drawings 2 and 3)*. After all the pilot holes have been drilled, ***the bit is replaced with one of 12mm diameter, and the speed reduced to not more than 400 rpm***. The holes are then re-drilled to the larger size, taking care to press only lightly on the drill-press handle. Pressing heavily when the second drilling is done will cause the drill bit to grab at the work. The four holes on the centre-line are ***not*** drilled to 12mm, but remain at 6.5mm.

When producing pumps in series, as many as eight valve plates can be clamped in the jig, and with the jig clamped in a machine vice and mounted on the drill press, a 6.5mm drill is used to drill the pilot holes directly, using the guide holes in the jig plate to position the drill accurately. After drilling through all the holes in the jig plate, remove the two nuts which attach the jig plate and hold-down bar to the jig plate, and replace the hold-down bar (without the jig plate) back on to the jig base. Ensure that the eight valve plates are properly aligned by placing 6mm diameter bolts into the four holes which lie along the centre line. Tighten the hold-down bar onto the valve plates and drill all of the holes in the valve plates to 12mm diameter (except the four centre holes in which the 6mm bolts have been inserted: these remain at 6.5mm diameter).

When the drilling is complete, the valve-plate is removed from the block of wood *(Figure 9, Drawing 5)*, and the lower surface is filed with a medium grade flat file to remove burrs of sharp metal caused by the drilling. It is important that this is carefully done, so as to avoid the risk that valve rubbers will be cut when the pump is in use.

The valve-box is assembled using an assembly jig provided with the basic tool kit. The jig is mounted in the steel vice, using the same mounting system used to hold the cylinder rolling tool in place. *(Figure 10, Drawings 1a and 1b)*. Set into the top of this jig are two threaded studs of 12mm diameter, spaced at the exact distance apart as the middle holes on the inner line of each set of port holes in the valve plate.

The valve plate is fitted over the threaded studs *(Figure 10, Drawing 2)*. A heavy-duty clamping bar with two 12.5mm holes pre-drilled (supplied with the toolkit) is set over the studs *(Figure 10, Drawing 3)*, and held in place with nuts *(Figure 10, Drawing 4)* which clamp the valve plate onto the jig. The edges of the valve plate must be flush with the edge of the mounting blocks on which it rests, and it may be necessary to flip the valve plate onto its other face, to accommodate minor inaccuracies in the position of the port holes. If the valve plate is badly misaligned this means that errors were made when the marking jig was used, or during the drilling of the port holes.

The next part to install is the side band of the valve box. ***The side band is cut from 3mm thick mild steel sheet, and measures 63mm wide by 670mm long. In countries where British dimensions are used the band can be made from 2½" by ⅛" steel strip***. If the band is cut from sheet it is very important to measure and cut the band accurately, so that its edges are parallel and straight. If this is not done the valve box is likely to leak where it contacts the baseboard. It may be practical to sub-contract the cutting of this (and other parts) to industrial manufacturers who own guillotine shears able to make straight, clean cuts in the steel sheet.

The side band is now welded to the valve plate. To do this, first clamp the side band to the valve box in the position shown *(Figure 11, Drawing 1)*. The bottom edge of the side band must lie flat against the base of the assembly jig, while the end of the band to be first attached to the valve plate should be about 5mm from the inside edge of the steel support block. It is best to clamp the side band to the outlet (discharge) side of the pump, because it is less important that this side be completely leak proof. The discharge side of the pump is that on which the 6.5mm diameter holes along the central axis are closest to the port holes. The side band can be clamped to the support block of the assembly jig with a 200mm (8") or larger C clamp. Securely tack-weld the side band to the valve plate where the two parts touch *(Figure 11, Drawing 2)*. Confirm before welding that the top of the side band and the valve plate are approximately the same height. ***Set the welding transformer to at least 100 amps, and use a 3mm (⅛") electrode.***

Without removing the C clamp, bend the side band around the first curved corner of the valve plate *(Figure 11, Drawing 3)*, making sure that the

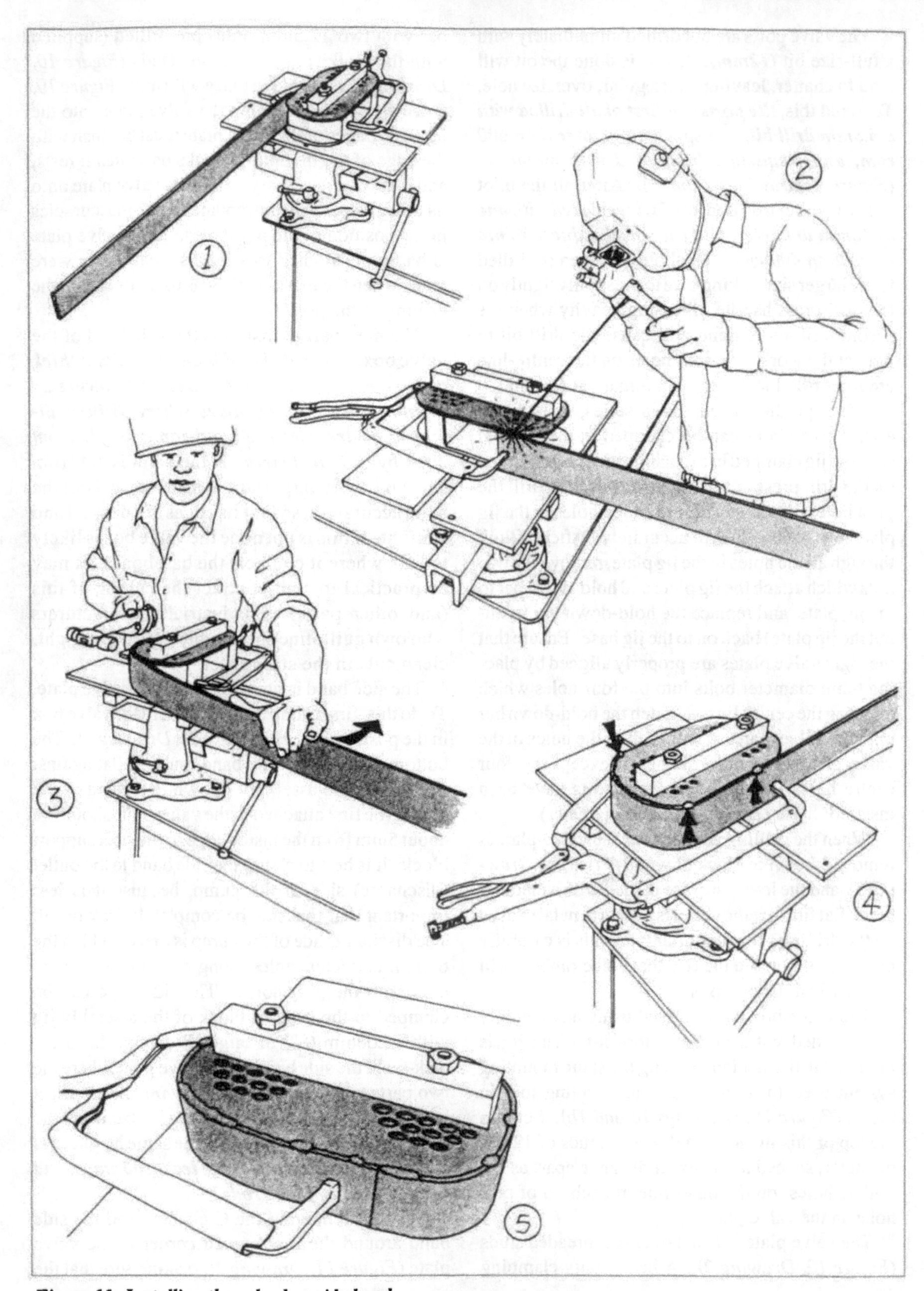


***Figure 11: Installing the valve box side band***

side band remains in contact with the base of the assembly jig. Tack the band to the valve plate once again, near the end of the valve plate between the first and second corners. Tacking the side band to the valve plate is much easier if two people work together, one bending the metal band and holding it against the valve plate with a steel bar (to avoid being burned), while the other tacks the band in position. It may also be occasionally necessary to tap the side band lightly with a hammer so as to close the seam between the side band and the valve plate. *(Figure 11, Drawing 3)*.

Remove the C clamp and bend the side band around the second corner of the valve plate. Use the clamp to tighten the band against the flat edge of the valve plate at one of the steel blocks of the assembly jig *(Figure 11, Drawing 4)*, which prevents the side band being bent inwards, and, where necessary, tap its upper edge towards the edge of the valve plate, and continue tack-welding. Continue bending the side band around the third and fourth corners of the valve plate in the same way, while tacking the band to the valve plate as before.

The ends of the side band should come closely together when the bending is complete. If the ends overlap slightly the end of the side band that has not yet been tack-welded must be trimmed. This can be done by removing the valve box from the assembly jig, and cutting off the excess length with a hack-saw. Replace the valve box on the assembly jig, clamp it down as before, and finish tacking the side band to the valve plate. Note the exact length of the side band that has been trimmed off so that future bands can be cut to the correct length before tacking them to the valve plate, thereby avoiding this trimming operation in the future. The nominal length of the side band is 670mm, but this will vary slightly depending on the exact dimensions of the valve plate template used. Always use the same template, which should be marked and set apart, so that the length of the side band need never be changed.

The next step is to make and attach the six pockets or bushings that hold the baseboard attachment bolts. ***Workshops that have a metal lathe can use bushings made from a 14mm diameter round mild steel rod. The rod is cut into lengths of 30mm and drilled lengthwise in the lathe with an 8.5mm drill. If a lathe is not used, pockets can be made by cutting 30 x 30mm square blanks from a 3mm thick sheet*** *(Figure 12, Drawing 1)*. These blanks are then bent in the vice to a semi-circular section *(Figure 12, Drawings 2,3 and 4)*.

Position the pockets (or bushings) for welding by loosely inserting six ***8mm x 70mm long bolts*** into the six holes bored into the baseplate of the assembly jig *(Figure 13, Drawing 1)*. The pockets are then placed over the bolts *(Figure 13, Drawing 2)* and held in place with a metal bar, pressed against the valve box

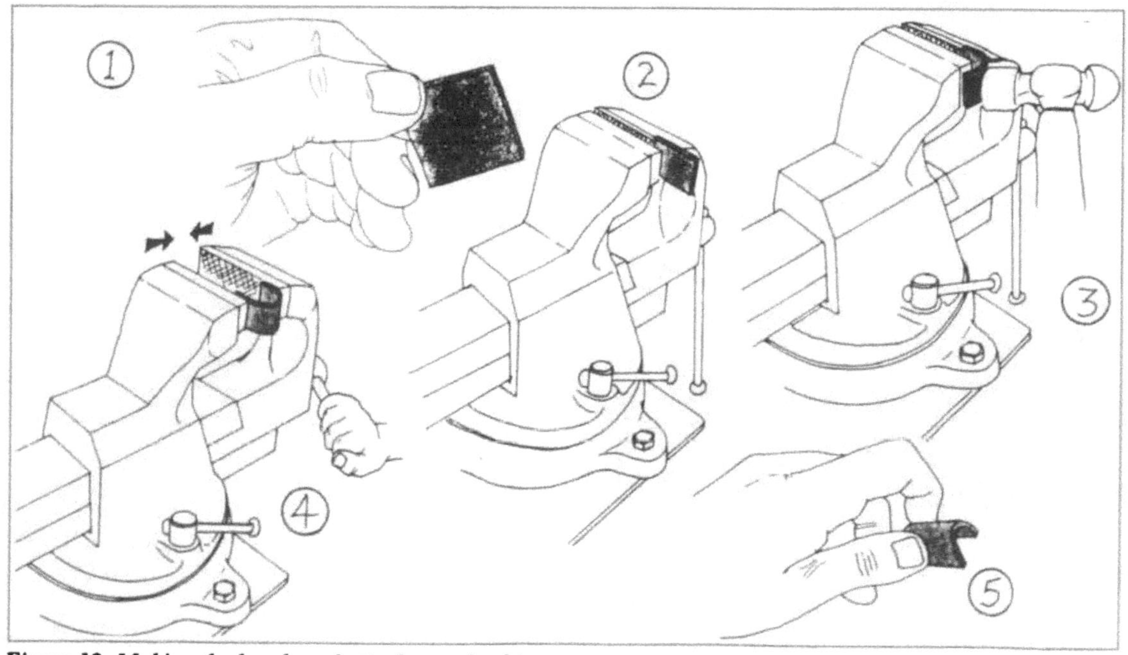


***Figure 12: Making the baseboard attachment bushings***

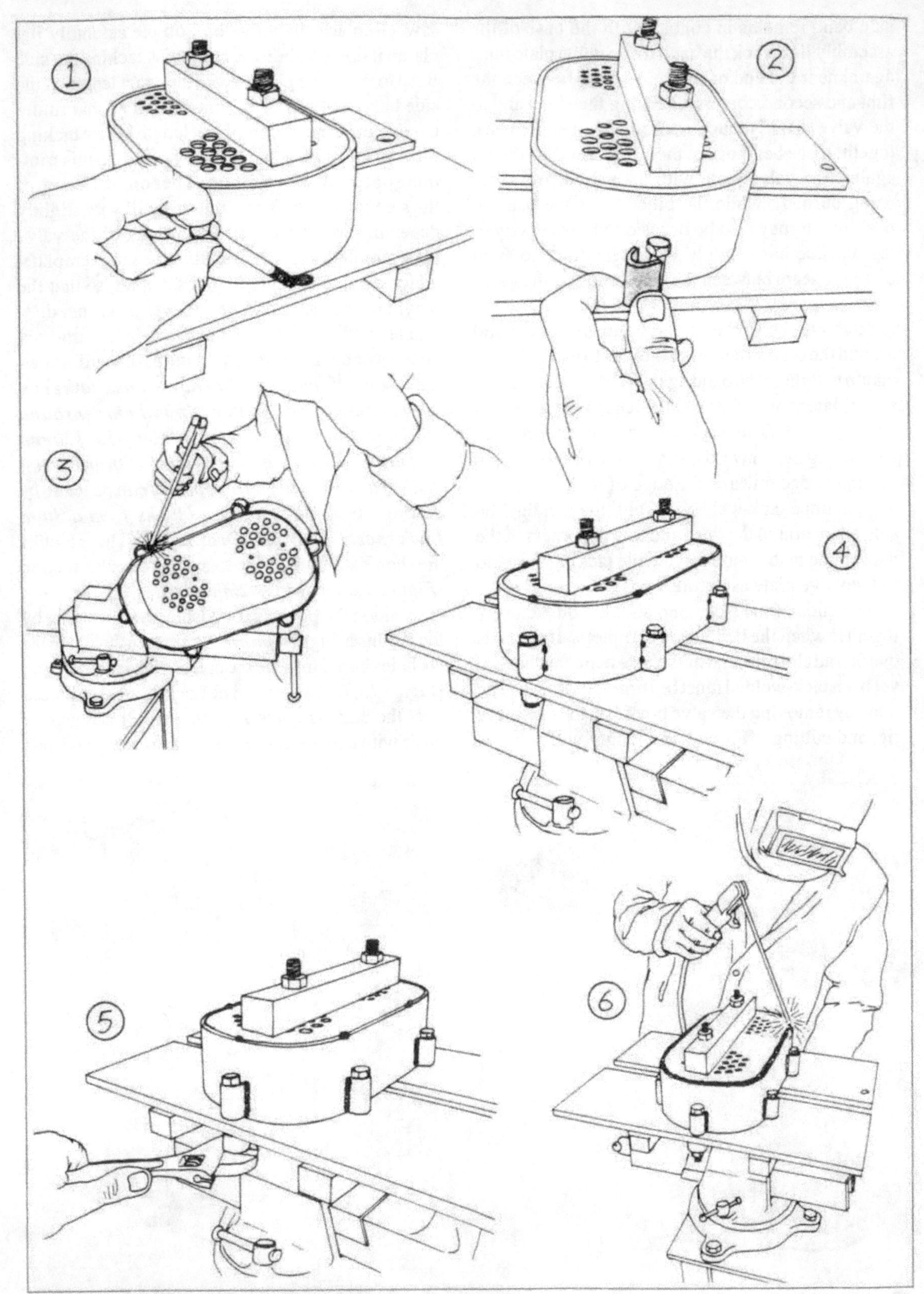

***Figure 13: Welding the pockets and final welding of side band***

body. The pockets are then tack-welded. Final welding of the pockets (or bushings) is best done after the valve box is removed from the jig, and held at a convenient angle for welding *(Figure 13, Drawing 3)*. It is important that the pockets are fully welded at this time, because the welding operations that follow are likely to distort the valve box unless it is securely bolted into place on the assembly jig: tack-welds are likely to separate under the stresses caused by this welding.

After securely welding the pockets (or bushes) into place, the valve box is put back on the assembly jig *(Figure 13, Drawing 4)* and the baseplate mounting bolts are re-inserted, and securely tightened into place *(Figure 13, Drawing 5)*. The heavy duty clamping bar is also re-attached. This bolting and clamping minimizes the possibility of distortions during the welding operations that follow.

Complete the welding of the side band, ***using a 3mm (1/8") electrode, with the welding transformer set at, or slightly above, 100 amps*** *(Figure 13, Drawing 6)*. The weld should not be too heavy, but strong and neat, to ensure that it is leak proof.

The next step is to cut the inlet and outlet holes in the valve box. The assembly jig has an attachment to guide a ***44mm diameter hole saw*** that is used to cut these holes. Four hole saws are provided with the toolkit. The hole saw is attached to a hand crank, which is slid into the slots provided on either side of the assembly jig *(Figure 14, Drawings 1 and 2)*. The slide is pre-lubricated with motor oil. A lever, which is also provided with the toolkit, is fitted into a hole in the baseplate of the assembly jig *(Figure 14, Drawing 3)*. The hand crank is rotated with the right hand, while pressure is applied on the lever with the left hand *(Figure 14, Drawing 4)*. It has been found practical to use an inner tube rubber band to apply pressure on the lever, instead of the left hand, leaving the operator free to choose the most comfortable operating position to turn the crank. A small amount of motor oil, or lard (white cooking fat) should be applied to the cutter to minimize friction, and extend the life of the cutter. With care, each hole saw should be able to cut several hundred holes before it is worn out.

The next step is to weld the cylinders to the valve box, and is one of the most difficult steps in producing the pump. First of all check that the base of each cylinder is regular, and use a set-square to check that the cylinder sits vertical to a horizontal surface when set on end. Use a flat file to correct any irregularities. Next, remove the clamping bar from the top of the valve box, but leave the base-board tie down bolts in place.

The cylinder is placed on the valve box in a position that is offset to the layout of the valve port holes *(Figure 15, Drawing 1)*. This is done to provide room inside the cylinders for seating of the inlet valves. Reference to *Figure 15, Drawing 1* shows the position of the cylinder when viewed from above the valve box. A 6mm steel disc (not illustrated) is supplied with the tool kit, and fits down over the assembly jig. It has two 6mm bolts which register in the 6.5mm holes along the centre of the valve plate. This disc has the same diameter as the inside of the cylinder and accurately positions the cylinder for welding. It is held in place by the nuts that normally secure the hold-down bar.

Welding the cylinders to the valve box requires some practice. This task is made more difficult by the fact that the cylinder material is thin and often slightly separated from the surface of the valve box by slight irregularities and warps in the valve plate, and cylinder ends. ***The welding should be done with a 3mm electrode and a current of between 90 – 100 amps***. The welding is started by lightly tack-welding the cylinders in place in three positions around the circumference, and checking that they are as close to perpendicular to the valve box as possible, with the smallest possible gaps between the cylinder bottoms and the valve plate. When finally welding the complete joint it can be done best if the electrode is pointed toward the cylinder (less than 45$^0$ above the horizontal), and if the tip of the electrode slightly leads the electrode holder *(Figure 15, Drawing 2)*. This position causes the electrode to throw molten metal from the valve plate towards the cylinder, and also it enables it to add metal without the risk of cutting or penetrating the cylinder.

If the cylinder to valve box welds are imperfect on the first pass of the electrode, ***they can be built up by raising the current to 105 amps*** and positioning the electrode at an angle even closer to the horizontal than on the first pass. Doing so will smooth the original weld, and seal any places between the original weld bead and the cylinder where leakage might have occurred. Imperfections on the inside of the cylinder caused by this weld will not affect the leather pump cups, because the

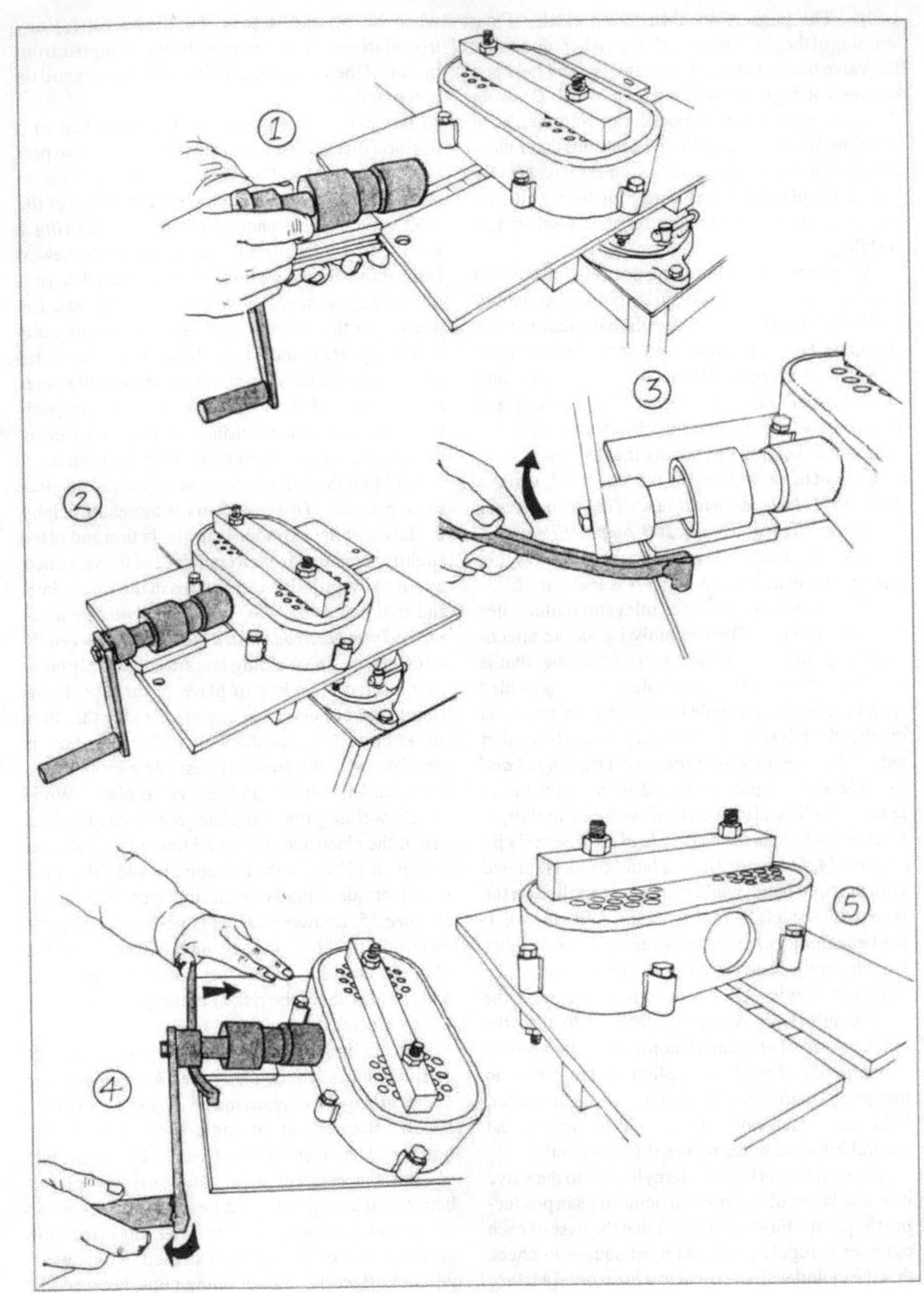


***Figure 14: Cutting the inlet and outlet holes in the valve box***

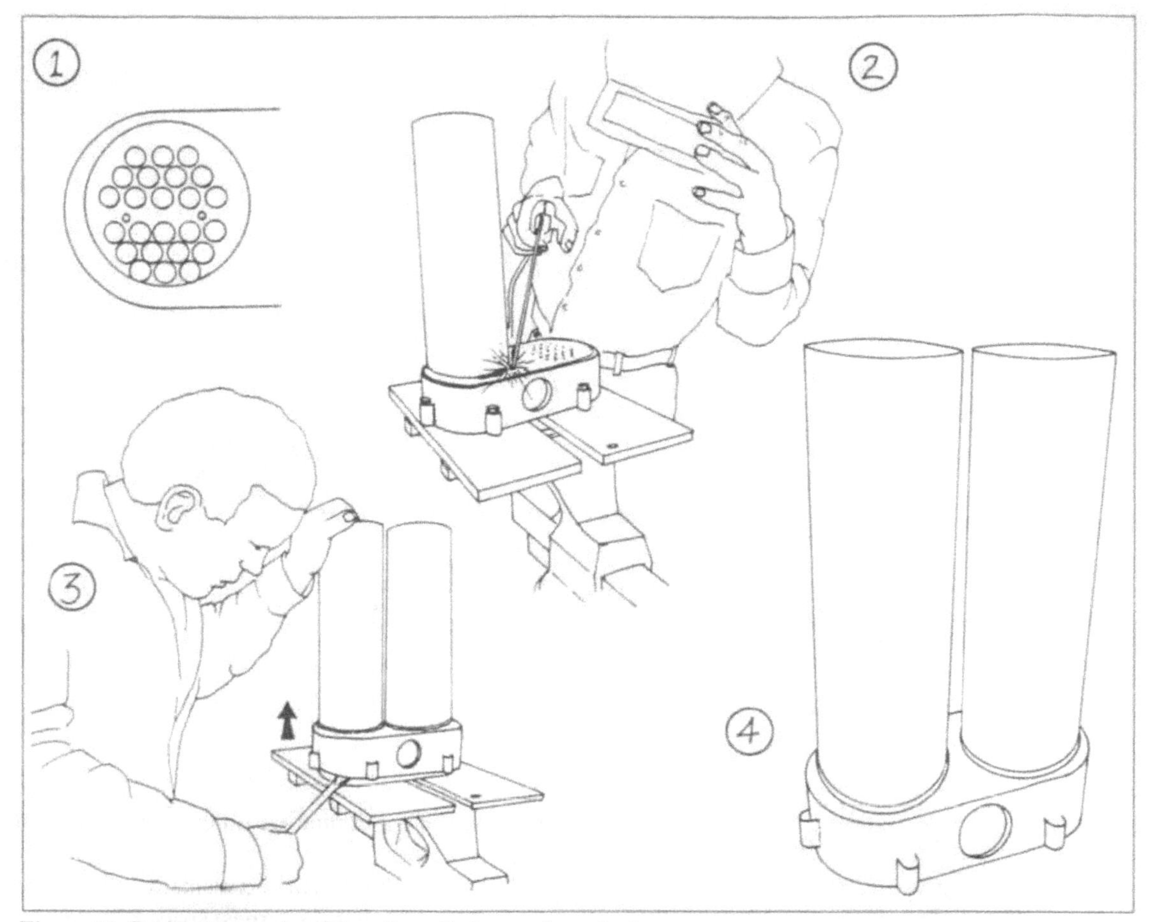


*Figure 15: Positioning and welding the cylinders to the valve box*

pistons are not permitted to reach the bottom of the cylinders on the downward stroke. However, it is good practice to make the welds as light as possible, consistent with a strong, airtight connection, to minimise warpage of the valve box and reduce the time, electrodes, and electricity consumed.

After completing the cylinder-to-valve box welds, the pump body may be removed from the assembly jig. To do this, first remove the 6 base-board tie down bolts. Prise the valve box off the jig by wedging a large, flat screwdriver under the valve box at the gaps in the assembly jig base, and then at positions around the edge *(Figure 15, Drawing 3)*. Be careful to lift the pump body vertically off the assembly jig. The completed pump body is shown *(Figure 15, Drawing 4)*.

The next step is to cut the inlet (suction) and outlet (pressure) pipes and weld them to the valve body. ***The inlet pipe and outlet pipes are cut from a length of 42 x 49mm black, ungalvanised water pipe. The inlet pipe is cut to a length of 900mm, while the outlet pipe is cut to a length of 100mm.*** Galvanized pipe can also be used but the zinc coating should be ground off where the pipe will be welded, to facilitate the welding, and to prevent the welder breathing noxious zinc oxide fumes. Grind or file a rounded edge, on the one end of each of these pipes that will ***not*** be welded to the pump body. This is done to facilitate the attachment of PVC couplings when installing the pump in the field.

The inlet and outlet pipes can be positioned for welding by inserting a length of round steel bar or pipe ***with an outside diameter of 40mm*** through the pump body *(Figure 16, Drawing 1)*. The two pipes are slid over this bar and butted against the valve box *(Figure 16, Drawing 2)*. Use of the round bar keeps the two pipes in line with each other, and with the holes in the valve box. Again, it is vital to ensure that the inlet pipe, which is the longer of the

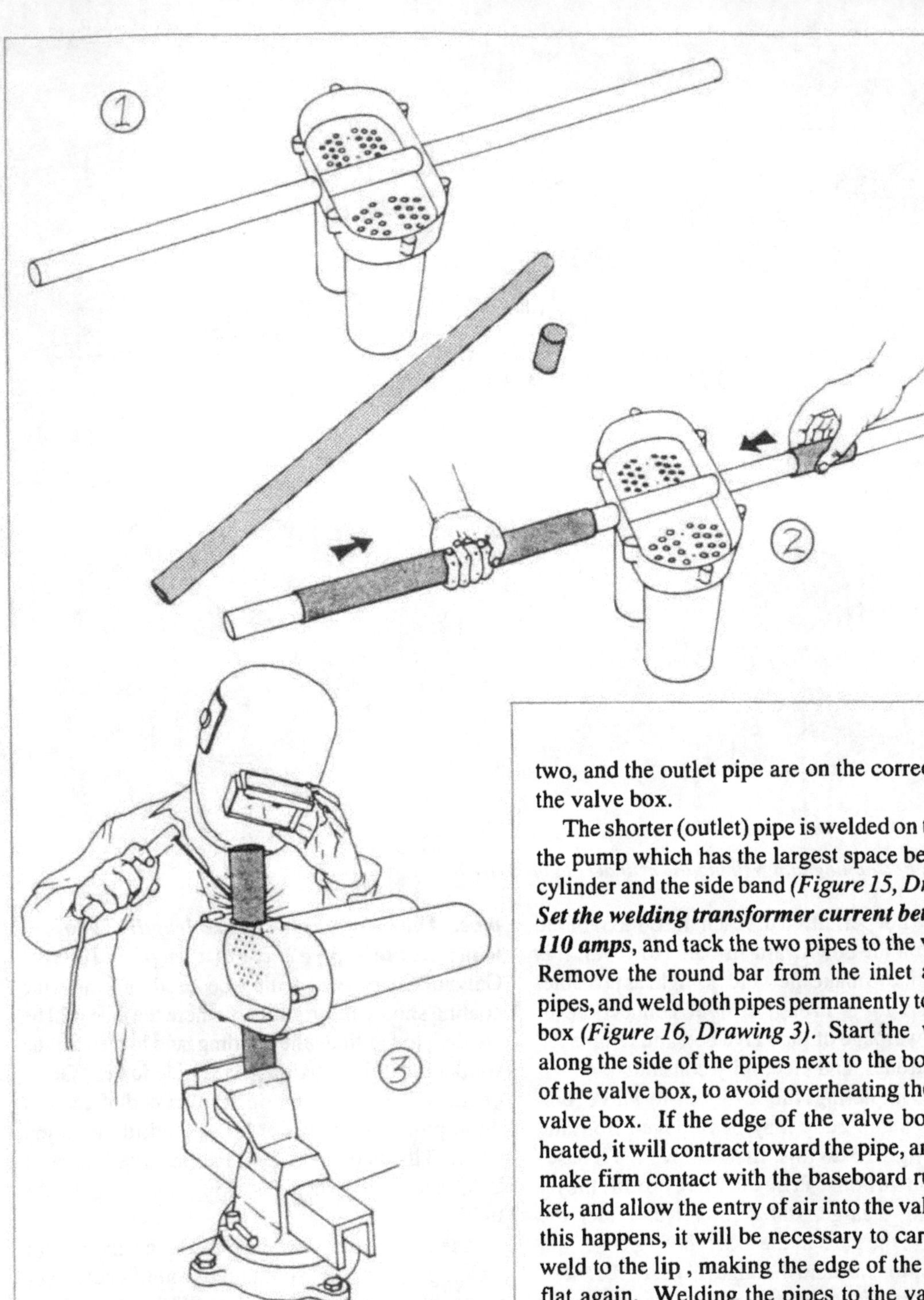


*Figure 16: Attaching the inlet and outlet pipes to the pump body*

two, and the outlet pipe are on the correct sides of the valve box.

The shorter (outlet) pipe is welded on the side of the pump which has the largest space between the cylinder and the side band *(Figure 15, Drawing 1)*. ***Set the welding transformer current between 90–110 amps***, and tack the two pipes to the valve box. Remove the round bar from the inlet and outlet pipes, and weld both pipes permanently to the valve box *(Figure 16, Drawing 3)*. Start the weld bead along the side of the pipes next to the bottom edge of the valve box, to avoid overheating the lip of the valve box. If the edge of the valve box is overheated, it will contract toward the pipe, and cease to make firm contact with the baseboard rubber gasket, and allow the entry of air into the valve box. If this happens, it will be necessary to carefully add weld to the lip , making the edge of the valve box flat again. Welding the pipes to the valve box is easier when the pump body is positioned so that the pipes are nearly vertical *(Figure 16, Drawing 3)*.

The final major step in making the pump body is to weld the valve box divider in place. First, try inserting the divider into the box *(Figure 17, Drawing 1)*. If it does not enter, carefully file or grind down

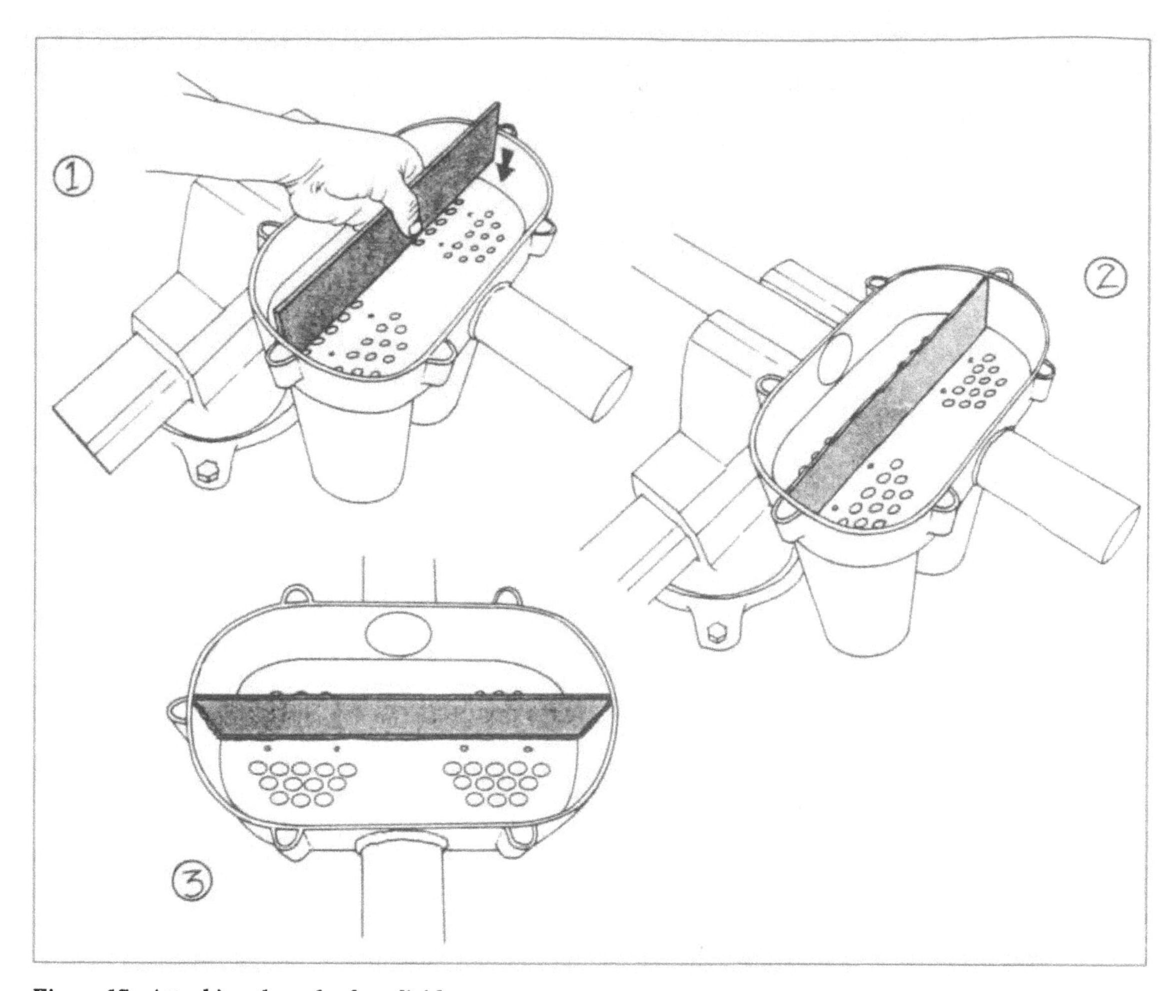


*Figure 17: Attaching the valve box divider*

both ends of the divider. Once the divider fits into the box, place it in its correct position, as far away from the valve rubber bolt holes, and as close to the inlet port holes as possible *(Figure 17, Drawing 2)*. This allows the maximum possible space for the valve rubber attachment bolts to be inserted during assembly. When the divider is in place, take a steel rule and check that the divider is at the same level as the bottom of the valve box along its entire length, so that the walls of the valve box ***and*** the divider will all make contact with the rubber gasket that will be positioned between the valve box and the base board during assembly. If the sealing edge of the divider is straight, but too high, file or grind the ***opposite side*** of the divider where it contacts the valve plate. When the divider is straight, and at the same height as the side band at both sides of the valve box, it can be tack-welded into position. Again, check that the divider is vertical and straight before continuing with the final weld.

The divider can now be permanently welded into place. ***Set the welding transformer between 70–90 amps, and use a 2.5mm electrode***. Make a smooth bead from the end of the divider towards the centre, but only on the side of the divider where the valve rubber bolt holes are to be found *Figure 17, Drawing 3)*. These welds are best made with the seam running slightly downhill, and the divider and valve plate set in such a position that they are at 45$^0$ to the vertical, permitting the welder to weld in a V trough. The bead must be smooth to avoid interference with the valve rubber tie down bolts.

The most difficult and time-consuming part of making the pump is now complete. With experienced workers building pumps in small batches (10–20) the above operations take about five hours to complete. Of this time, roughly one and a half hours are spent with the pump body mounted in the assembly jig. Because of this, no more than four or five pumps can be built per day in a workshop which has only one assembly jig. The assembly jig will, however, probably not be the factor limiting the rate of pump production, since roughly eight experienced workers would be needed to make that many pumps every day (two workers can make one pump per day between them). If the pump were produced in a large series in a large workshop, having a guillotine shear, metal lathe, and several arc welders, the production rate could rise to as much as one pump per person-day. At this rate, two assembly jigs would be required.

# 3. Making the handle and treadle support

***The vertical handle pipe, handle T, and treadle support pipe are all made from 21mm x 27mm black water pipe.*** Again, galvanized pipe can be used, but the zinc coating should be removed wherever the pipe will be welded. Pipe of this type has a 3mm wall thickness ***and should be welded with a 3mm electrode, at a current of at least 105 amps*** so that the welds are as strong as the pipe itself.

***The vertical handle pipe is cut to a length of 1350mm (52"), and the T is cut to a length of 400mm (16").*** The vertical pipe and handle T can be easily joined if the end of the vertical pipe is flattened in the vice, prior to welding to the centre of the T piece *(Figure 18, Drawings 1 and 2)*.

***The shaft (axle) around which the pulley rotates is made from smooth round rod, at least 18mm in diameter, and at least 50mm in length. It should have a 4mm diameter hole drilled through it about 5mm from the unwelded end to accommodate a cotter pin. This shaft is butt welded to the vertical shaft at a distance between 850–1 000mm from its lower end.*** Tack-weld the shaft to the handle pipe *(Figure 18, Drawing 3)*, and then make sure that it is square with respect to the pipe, and perpendicular to the T when the handle pipe is viewed endwise. Weld the pulley shaft in place, with a good strong fillet, bearing in mind that this weld will have to support the weight of two adults. File a chamfer on the end of the pulley shaft (or previously do so in a lathe), and remove any weld spatter or burns on the shaft. If the pulley shaft is not smooth, or less than 18mm in diameter it will quickly wear out the hardwood pulley when the pump is in use. If the manufacturer intends to use a commercially available cast iron pulley in place of the hardwood pulley, the shaft should correspond in diameter and length to the bore of the cast iron pulley. Cast iron pulleys are more durable than those made of hardwood, and are readily available in many countries, but tend to be costly.

The handle is attached to the pump body by two steel sleeves or bushings that are welded to the pump body, on the back (inlet) side of the pump. ***These are made from 30mm lengths of 26mm x***

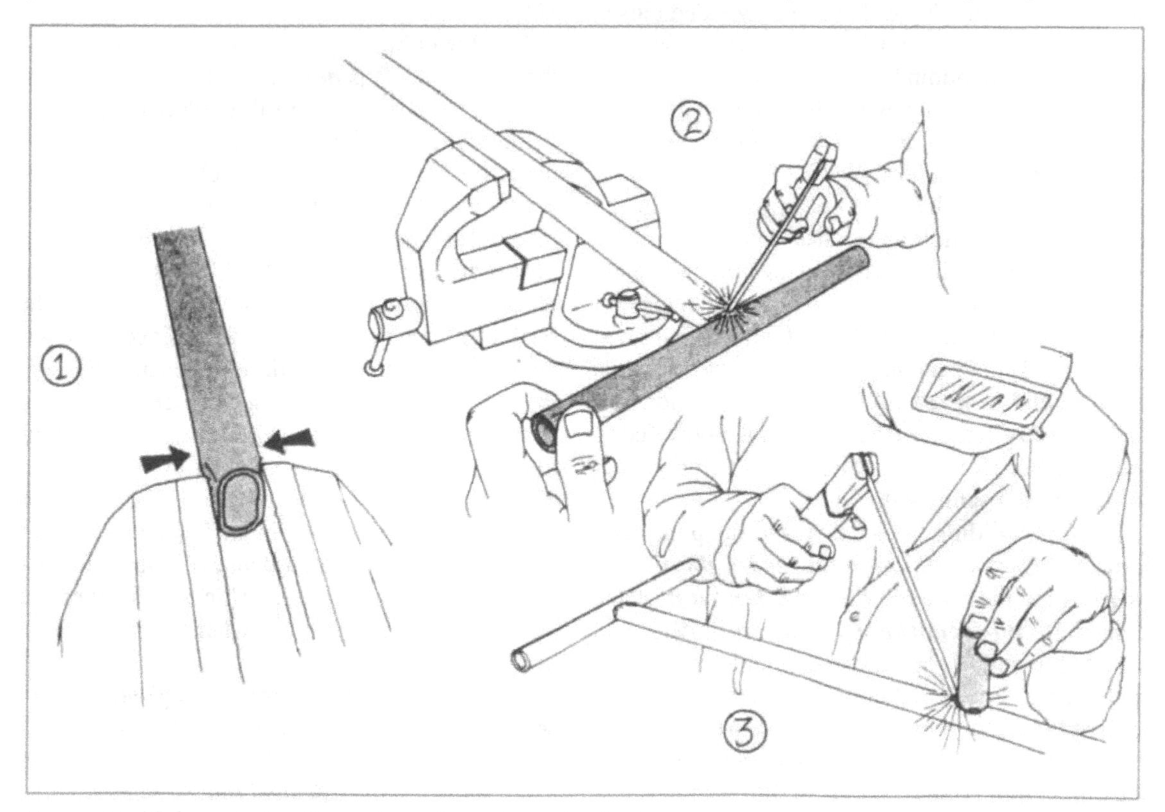


***Figure 18: Making the handle, and attaching the pulley shaft***

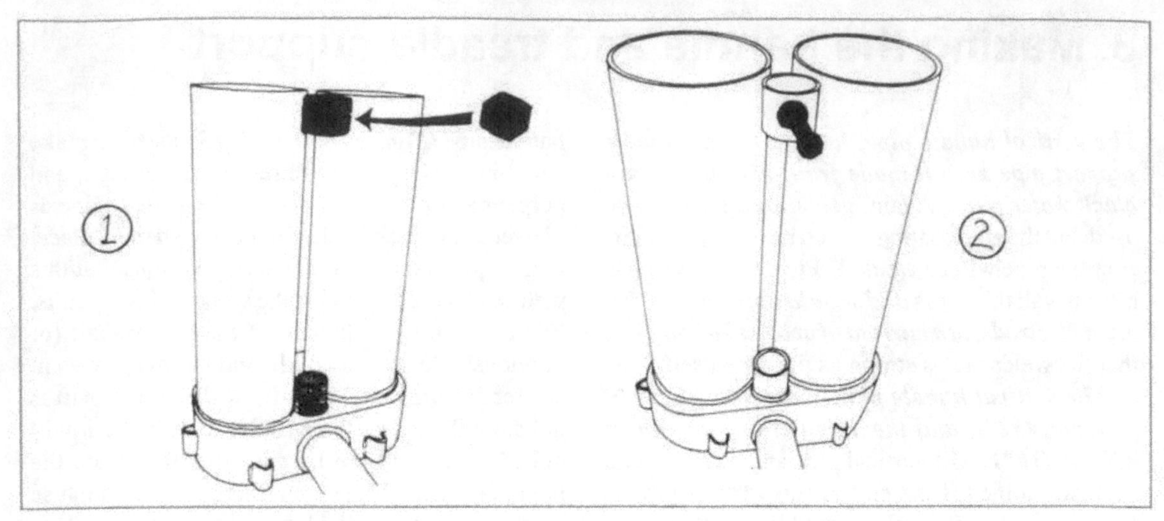


*Figure 19: Attaching the handle support bushings to the pump body*

***32mm diameter water pipe.*** The inside diameter of these sleeves must be increased to allow the handle to slide through. If a metal lathe is available, ***bore the sleeves to approximately 27.5mm or 0.5mm larger than the outside diameter of the handle pipe.*** If a metal lathe is not available, cut a slit down one side of the sleeve using a hacksaw. Enlarge the sleeve by prying the slit open with a flat screw-driver. The slits do not have to be welded closed because of the manner in which the sleeves are attached to the pump body.

The upper bushing, which is welded to the top of the pump cylinders, ***is drilled with an 11mm hole, and a 10mm nut welded over this hole.*** *(Figure 19, Drawing 1)*. A bolt should be threaded through the nut and into the hole in the sleeve while the nut is welded into position. This will ensure that the nut and the hole are properly aligned, and that the bolt presses on the handle shaft at 90$^0$ *(Figure 19, Drawing 2)*. The nut is best welded attached when both bushings have been welded into position (see below). The lower bushing does not have a fixing nut and bolt for the handle, and is therefore not drilled in the same way.

Before welding the bushings to the pump cylinders, ***insert a 600mm or longer length of $^3/_4$" water pipe into the top bushing, hold it in place about 10mm from the top of the cylinders***, and allow one end of the pipe to descend towards the valve box. Fit the lower bushing over the bottom end of the water pipe, and let it rest in place on top of the valve box. Press both upper and lower bushings lightly against both of the cylinders, and adjust the position of the lower bushing to ensure that the pipe is exactly parallel to each of the two cylinder walls when viewed from both directions. To locate the lower bushing correctly, it is usually necessary to grind or file two chamfers on one end of the bushing to make room for the welds which attach the cylinders to the valve box. Tack-weld both of the sleeves into position, and remove the 600mm pipe.

The lower bushing is welded only to the top of the valve box. It is ***not*** welded to the cylinders to minimise the risk of burn-through damage to the cylinders in a position where repair is difficult. The upper bushing is welded securely to the cylinder walls, and requires two or three passes of the electrode on each side of the sleeve ***using a 2.5mm electrode, and a current of 90 amps.*** Point the electrode towards the cylinder (away from the bushing) when making these welds since the cylinder is of lighter gauge. If the arc does burn through the cylinder wall, the damage can be repaired by filing the inside of the cylinder smooth. Burn-through can usually be avoided by moving the electrode quickly and steadily along the weld, and pointing it towards the cylinder. Secure welding of the upper bushing is particularly important, because it must resist considerable forces generated by pumping operations, and when operators climb on and off the pump.

***The treadle support pipe is made from a 600mm length of 21 x 27mm black pipe. The treadle shaft (T-piece) is made from a 220mm length of 12mm ($^1/_2$") diameter smooth steel rod.***

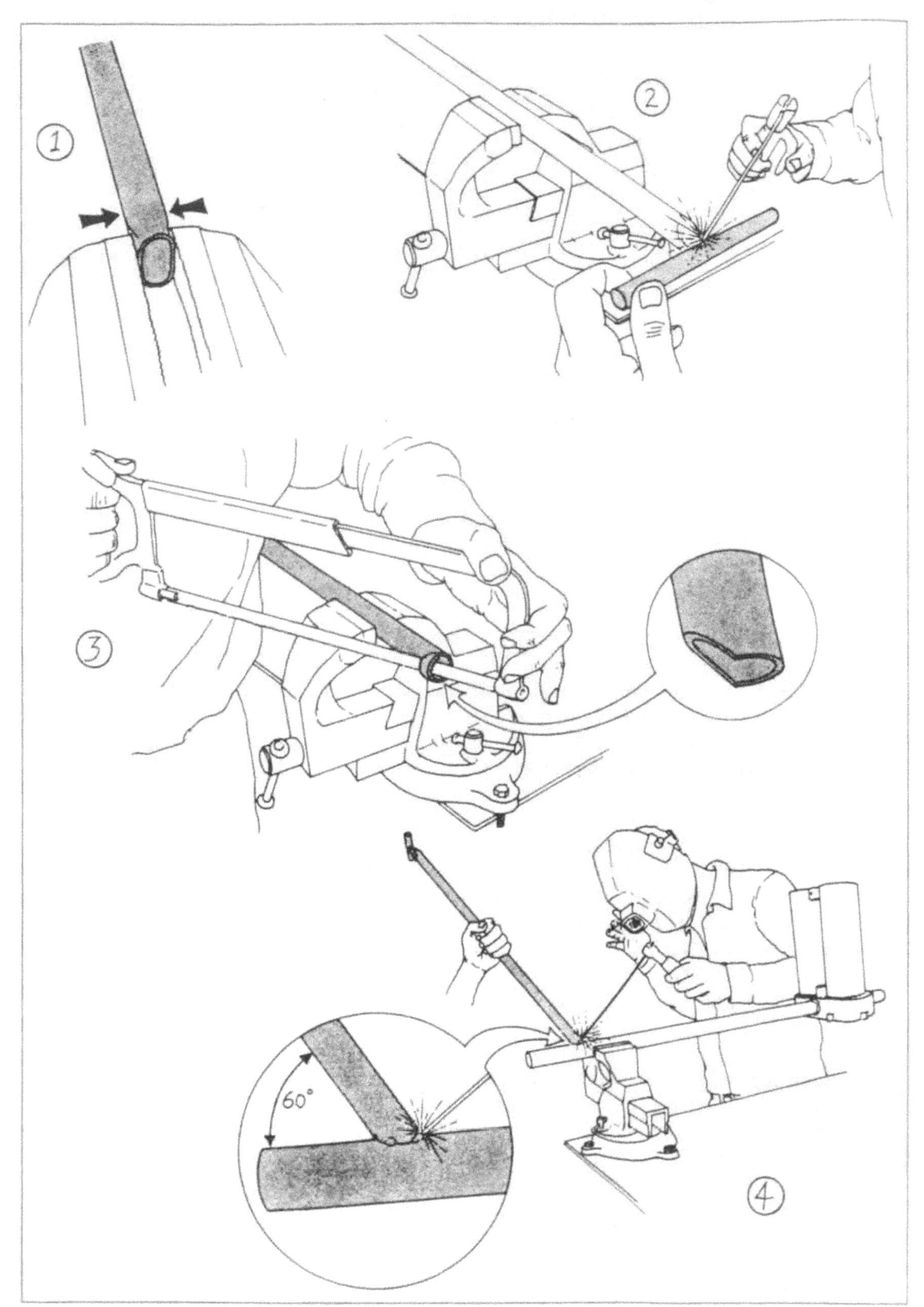


*Figure 20: Making and installing the treadle support*

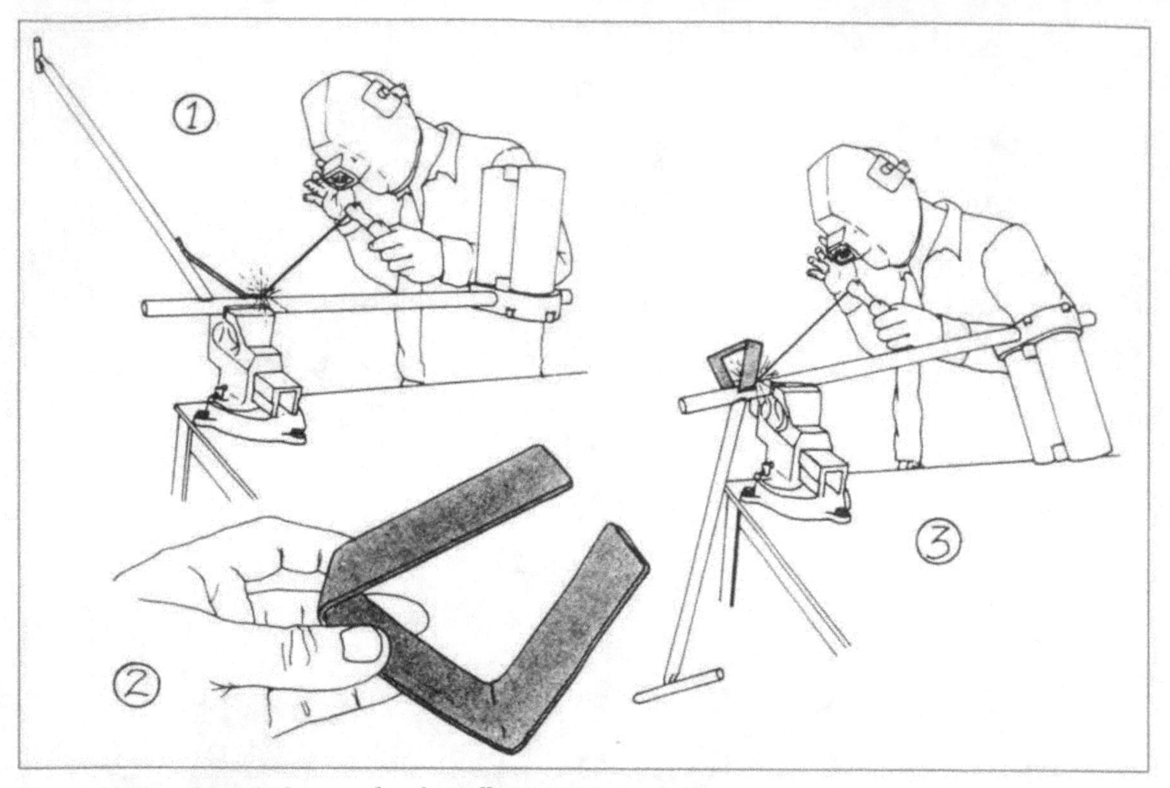


*Figure 21: Attaching the brace rod and treadle support toot*

***Drill a 4mm diameter hole through each end of the treadle shaft, approximately 5mm from the ends*** (for pins which retain the treadles in place) Then, flatten one end of the treadle support pipe in a vice *(Figure 20, Drawing 1)*, and weld it securely to the middle of the treadle support shaft *(Figure 20, Drawing 2)*. The lower end of the treadle support pipe is then cut and filed *(Figure 20, Drawing 3)*, so that it fits snugly around the water inlet pipe when tipped at an angle of $60^0$ to the intake pipe *(Figure 20, Drawing 3 – inset)*. Tack-weld the lower end of the treadle support pipe at a distance of 100mm from the end of the intake pipe. Confirm that:

- the treadle support pipe is at the proper distance from the end of the intake pipe and set at the proper angle[18]
- when viewed from the intake end, the treadle support pipe is parallel to the pump cylinders
- when viewed from above, the treadle shaft is perpendicular to the inlet pipe.

Securely weld the treadle support pipe to the intake pipe, and remove the welding jig (if used).

Then bend the treadle support brace rod, made from 8mm diameter round bar, and weld it to the treadle support, and intake pipe *(Figure 21, Drawing 1)*.

***The treadle support foot is made from a 63mm ($2^1/_2$") strip of 3mm ($^1/_8$") thick stock, or similar material.*** The foot is bent round a form that is included in the toolkit. Position the strip so that its two ends extend equally beyond the ends of the form and tighten the strip and the form in the vice. Then, using a medium hammer, bend both ends of the strip until they contact both sides of the form. When the foot is complete *(Figure 21, Drawing 2)*, position the foot under the intake pipe and weld it securely to the pipe *(Figure 21, Drawing 3)*.

The pump body is now completely finished *(Figure 22, Drawings 1a and 1b)* and ready for painting. Before painting it, make sure that all of the welds are cleaned of slag and spatter and that there are no faulty welds in the pump body that

---

[18] This jig is not provided as part of the tool kit.

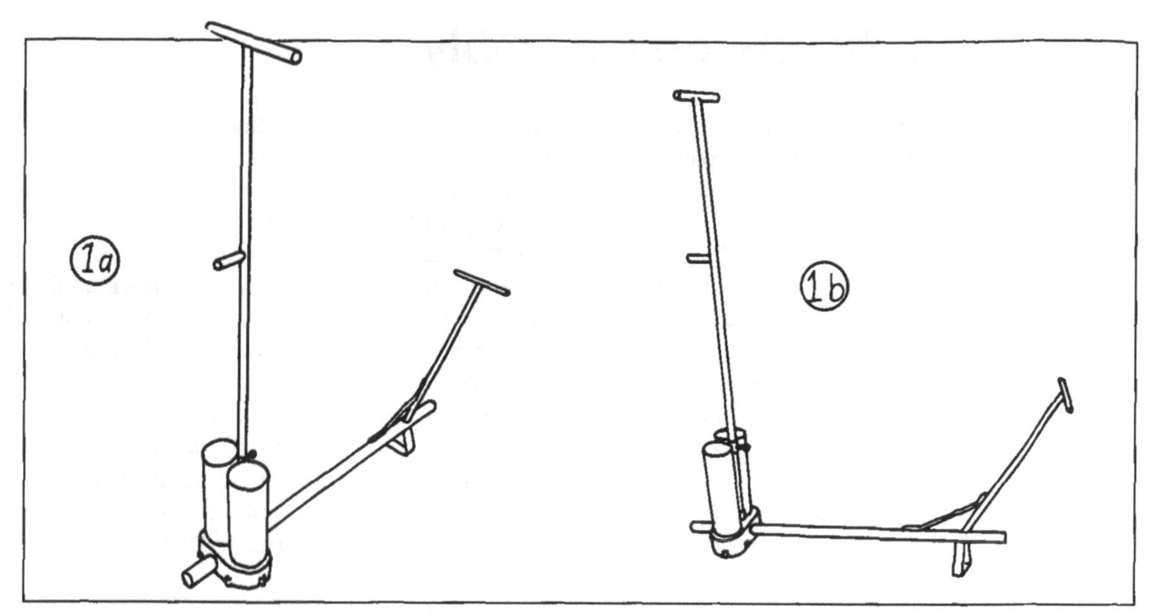


***Figure 22: Pump body complete with handle and treadle support***

could cause it to leak. Leaks at the cylinder seams and the cylinder to valve box welds are particularly serious because they enable air to leak into the pump, and prevent it from drawing water when working in suction. Finally, check that the bottom of the valve box is flat, where it will contact the baseboard, siting across the bottom of the valve box to re-confirm this.

When all of these points have been verified, clean the entire surface of the pump body with kerosine (paraffin), or another solvent, and paint it with anti-rust paint (red oxide). ***Leave the insides of the cylinders unpainted.*** When the anti-rust paint has completely dried, paint all surfaces except the inside of the cylinders with enamel paint. The insides of the cylinders are not subject to rusting because they are lubricated and polished by the leather pump cups. Painting them will foul the pump cups, and make pumping very difficult.

When producing pumps in small series, the above operations require approximately two man-hours per pump. The time could be reduced to a little over one hour when producing large numbers of pumps in an efficiently organized workplace.

# 4. Making the piston assembly

The pistons of the treadle pump consist of two leather cups, sandwiched between a pair of steel washers, all of which are held together by a nut and bolt welded to the end of the piston rod. A third steel washer of a larger diameter is placed between the two leather pump cups. The upper and lower washers serve to stiffen the pump cups, while the intermediate washer, made to a size only 2mm smaller than the pump cylinders helps to centralize the piston in the cylinder, and support the walls of the pump cups when the pump is in operation.

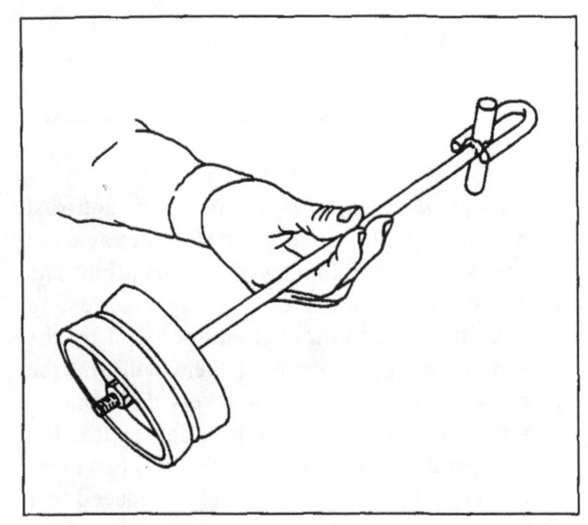

*Figure 23: Complete piston, piston rod and pump cup assembly*

The upper pump cup faces towards the top of the cylinder. It seals the piston when the pump is acting on the suction stroke, when the piston is moving upward. The lower cup faces towards the bottom of the cylinder. It seals the piston when the pump is acting on the pressure stroke, when the piston is moving downward. Because the leather cups are flexible, they expand against the cylinder as a result of atmospheric pressure (for the upper cup), and the pressure of the water inside the cylinder (for the lower cup), thereby sealing the piston.

***The pump cup support washers are made from 3mm thick mild steel sheet. The upper and lower washers are 90mm in diameter, while the intermediate washer is 100mm in diameter. The washers have a 12.5mm (1/2") centre hole.***

To make the smaller of the pump cup support washers, first trace parallel lines 47mm apart, across one side of a 3mm thick steel sheet. Then, trace parallel lines, perpendicular to these 47mm from one edge of the sheet, and every 47mm thereafter. Cut the sheet into strips 94mm wide along every other line of the first set of lines. Straighten these strips and centre punch every other intersection of lines, corresponding to the centres of the washers to be made. Using a metal compass, scribe circles around these punch marks, 90mm in diameter if the circles are to be cut directly on the shear, or 94mm in diameter if the disks are to be finished on the metal lathe. Drill the 12.5mm centre hole for each washer on the punch marks, and cut the strips into squares measuring 94mm on a side along the scribed lines. If the washer will be cut directly to their finished size using the shear, make a continuous cut around each washer, rotating the washer while pulling the handle of the shear. If the washers are to be cut on a lathe, the accuracy and smoothness of the cut are less important, but a continuous cut will save time, and reduce the risk of tool damage.

For every two disks which are cut in this way, to a diameter of 94mm, another disk should be prepared with a rough diameter of 104mm, making a set of three for each piston assembly. In this case, the scribed lines on the 3mm thick mild steel sheet should be 52mm apart, in both directions.

When at least 10 disks have been cut in this way they are mounted together on a mandrel made from a 12mm bolt, and turned to their respective sizes on a lathe *(Figure 24, Drawing 1)*. The smaller disks will be 90mm in diameter, and the larger disks will be 100mm in diameter. Each pump requires four of the smaller disks, and two of the larger. This same illustration shows how the disks are set on to the finished piston rod, and show that the upper and lower disks have curved edges facing towards the pump cups *(Figure 24, Drawings 2 and 3)*. This helps to prevent damage to the pump cups from sharp edges to the washers. This curved edge can either be prepared by manually filing each washer, or by individually turning the curved chamfer in another smaller mandrel. The illustration shows a group of 10 washers being turned on an unsupported mandrel. If more than 10 washers are to be prepared, the mandrel should be centre-drilled, and supported by the tailstock of the lathe.

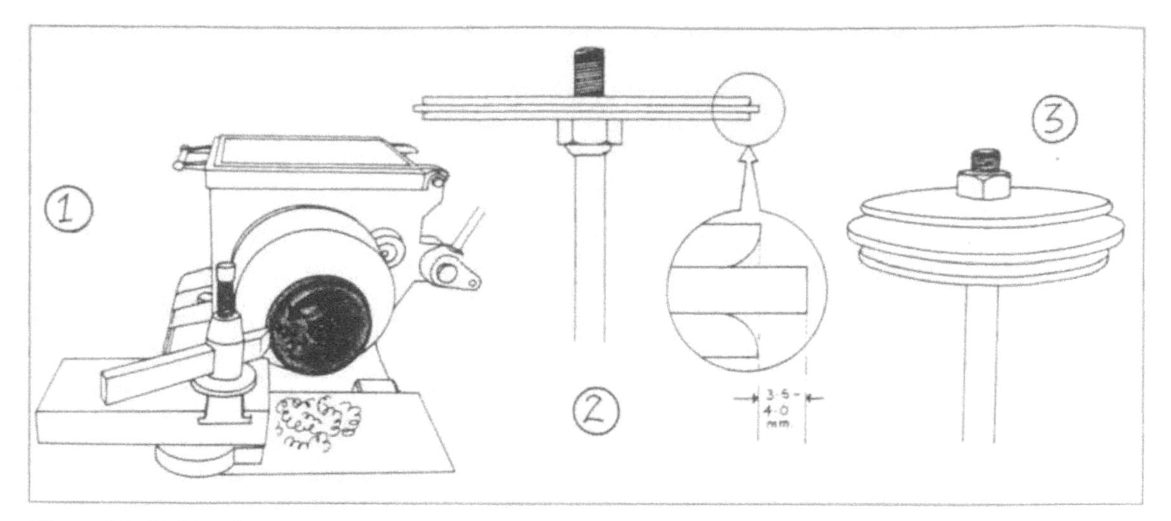


*Figure 24: Making the pump cup support washers*

***The piston rods are made by welding a 300mm length of 12mm diameter round rod to the head of a 12mm diameter by 25mm long bolt.*** First of all clamp a suitable nut into the vice, and screw the bolt into the nut, making sure that the nut is in a vertical position *(Figure 25, Drawing 1)*. ***Using a 3mm electrode, set the transformer to 105 amps*** and weld the bolt securely to the piston rod *(Figure 25, Drawing 2 and inset)*. It is important to use round rod, and not building reinforcing bar of a similar diameter. If reinforcing bar is used the piston rods will bend when the pump is used, and the welds may break owing to the higher carbon content of the steel used in reinforcing bar. ***The crosspieces are made from 80mm lengths of the same roundbar (12mm diameter), and welded to the piston rod using a 3mm electrode, with the transformer set to 120 amps*** *(Figure 25 drawing 3)*.

***The U piece of the pump rod which is used to attach the pulley rope, is made from a 200mm length of 8mm round rod***, bent in a vice as shown *(Figure 26, Drawings 1,2 and 3)*. It is easier to make this piece by bending the ends of a long piece of rod and cutting off this part so that its ends are of

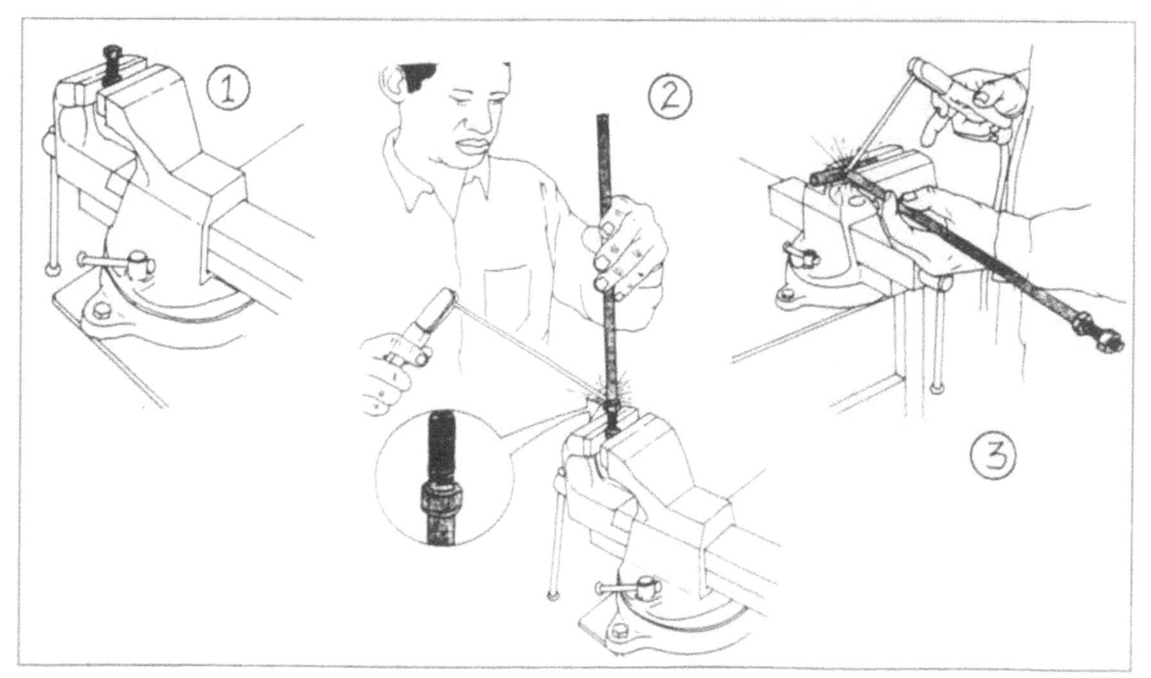


*Figure 25: Attaching the piston mounting bolt and T- piece to the piston rod*

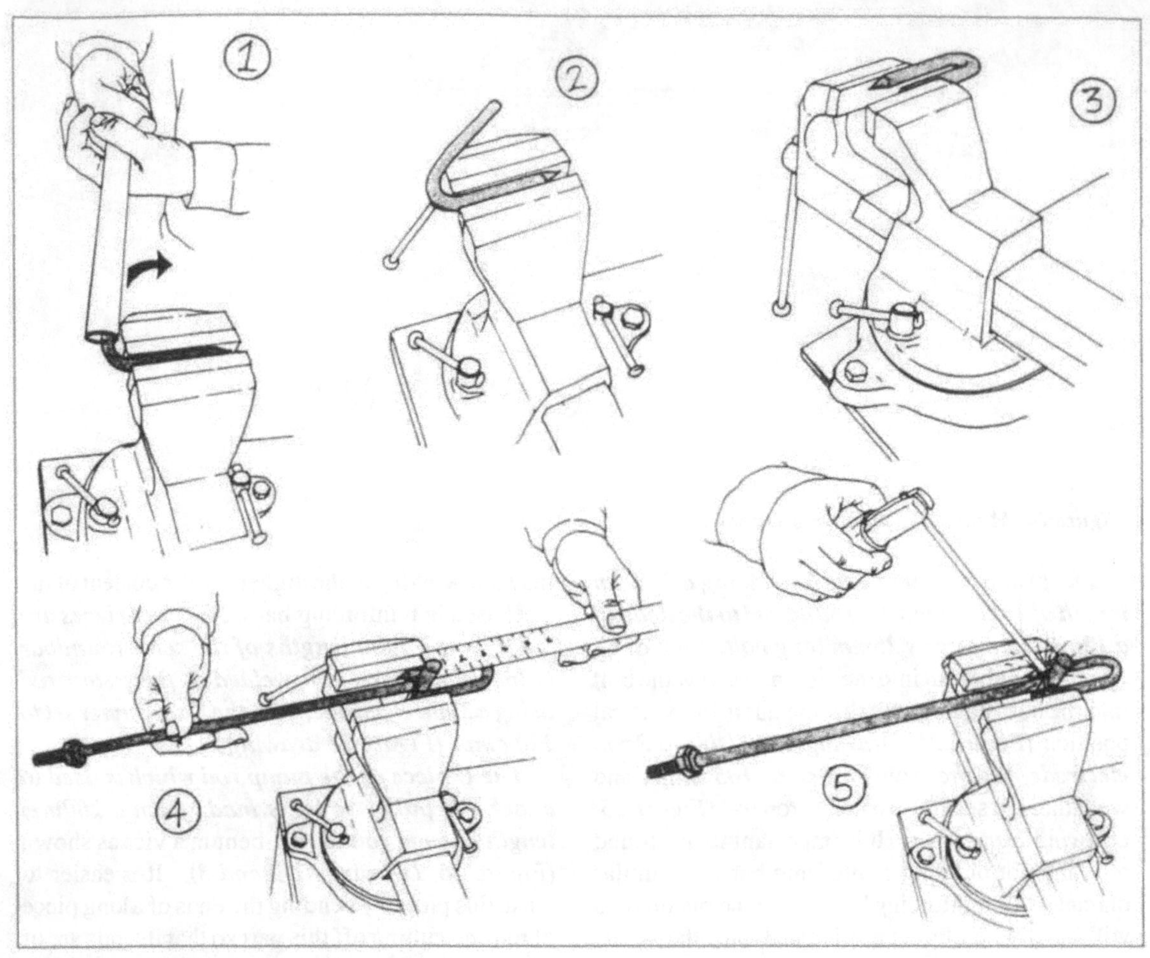


*Figure 26: Attaching the U piece to the piston rod*

equal length. This avoids the problem of cutting the rod to length first and then bending it, only to find that the ends are unequal. However, the illustration shows this part being made from 8" long nails, with the head removed, and if these are available they are usually cheaper than buying round rod.

Tack-weld the U piece to the piston rod and make sure that it is square with the crosspiece, when viewed from the upper end of the piston rod. ***The distance between the T piece of the piston rod and the top inner edge of the U piece should be not less than 60mm (Figure 26, Drawing 4)***, although this is not a critical dimension. ***Weld the U piece to the piston rod with a 3mm electrode, and the welding transformer set to 120 amps (Figure 26, Drawing 5).***

***The leather pump cups are made from 130mm diameter disks of good quality tanned cowhide, having a thickness of between 4 and 5mm.*** The type of leather used in making the pump cups is as important as its thickness. Leather processed for making shoe soles is too stiff, does not absorb water easily, and should ***not*** be used. This leather usually has a shiny surface, and is very hard and inflexible. Leathers that have been dyed, or are very soft, such as those used in making bags and leather clothing are also unsuitable. The best leather is natural in colour (pink to light brown), of medium stiffness and has a minimum of cracks, holes, or other defects. It is the type of leather usually used for making belts, saddles, and shoe uppers.

ATI and CARE have both experienced considerable difficulty in identifying and enabling pump makers to procure good quality leather. While leather tanned by traditional methods[19] in Africa has been used with success in the pump, the life of such leathers is measured in weeks, rather than seasons or years. Wherever possible, leather should be properly tanned, and can often be obtained from large-scale shoe factories, where it is prepared on specialized machines to an exact thickness.

Where whole hides are used, the pump maker should be aware that a hide varies in thickness, being thinner towards the belly, and some of this thinner leather may not be usable. By carefully drawing circles on the leather *(Figure 27, Drawings 1 and 2)*, at least 50 disks, or enough to make 12 pumps, can usually be made from a half-hide. It is best to draw the circles with a pencil so that they can be moved around as required to get the biggest number from the piece of leather. Another method is to place disks that have already been cut onto the remaining leather, and to move these around until the maximum number of usable disks can be obtained from the hide. Remember that parts of the hide that are cracked, have a hole, are badly wrinkled, or are too thin should not be used, and should be excluded when planning how to cut the leather.

The leather disks can be cut using heavy-duty scissors, or with a heavy-duty carpet knife. ***This can be facilitated by using a metal disk of 130mm diameter as a template*** *(Figure 27, Drawing 3)*. Before moulding the leather disks, punch a 12mm diameter hole in the centre of each disk. The centre is determined by carefully locating the centre hole of a template to trace the centre of the disks *(Figure 27, Drawings 4 and 5)*. When the hole position is properly marked it is punched by using a specially made punch *(Figure 27, Drawing 6)*. ***This punch is made from a piece of heavy duty steel tube, 120mm long and with a bore of 12mm.***[20] One end of the punch is chamfered. The leather disk is placed on a block of hard wood and the punch is carefully located over the centre mark. The punch is then struck with a hammer, which cuts the centre hole in one motion *Figure 27, Drawing 7)*. The piece of leather removed is forced up the inside of the hollow punch, and can later be removed with a wire. It is important that this is done after every two or three holes have been punched, to avoid clogging the bore.

---

[19] Traditional methods of curing leather involve the lavish use of salt, and sun drying. When used in the pump such leathers become over-penetrated by water, and soggy.

[20] Neither the template used to mark out the disks and the centre hole, nor the hole punch are supplied as part of the basic tool kit.

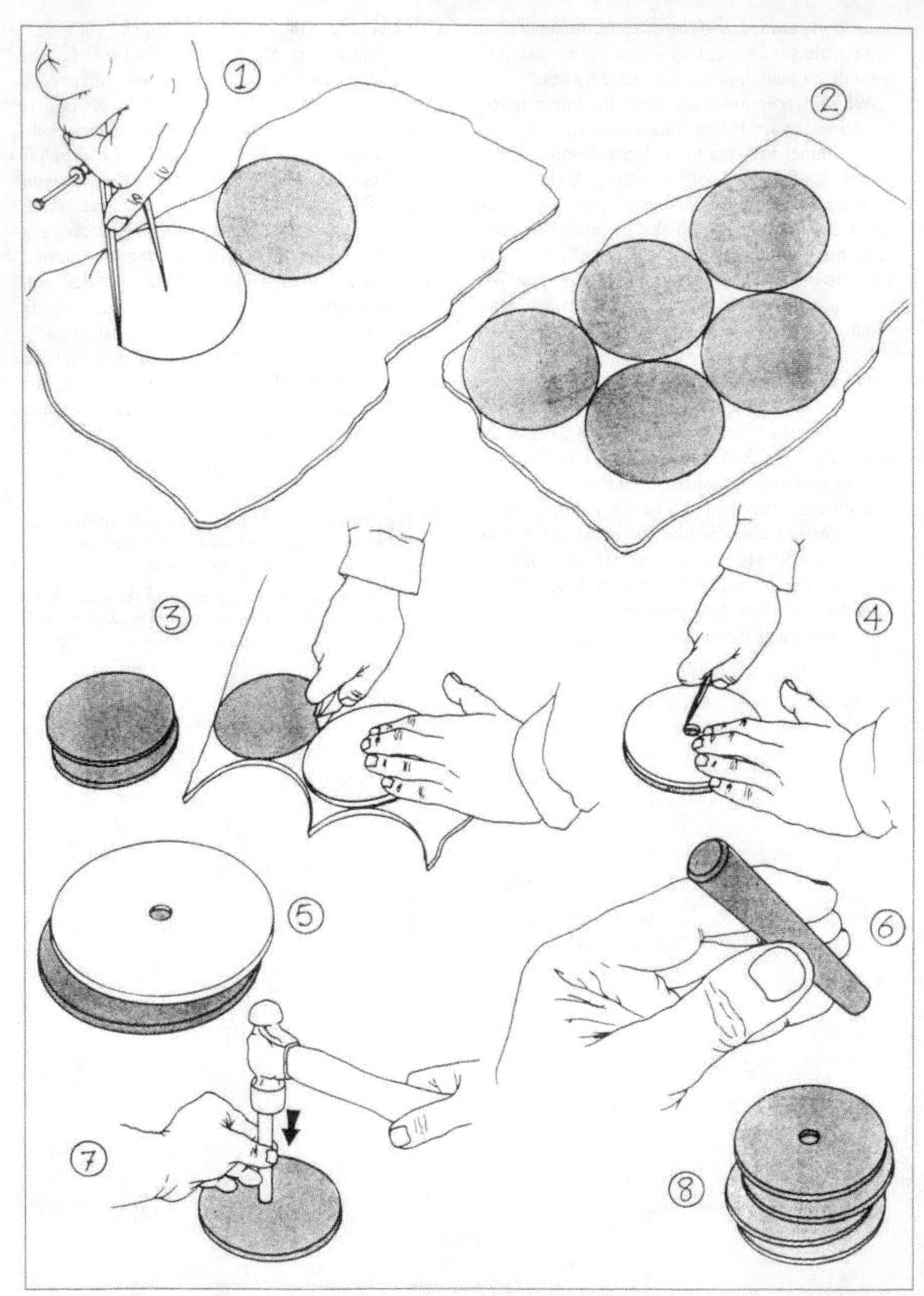


***Figure 27: Making the leather pump cup disks***

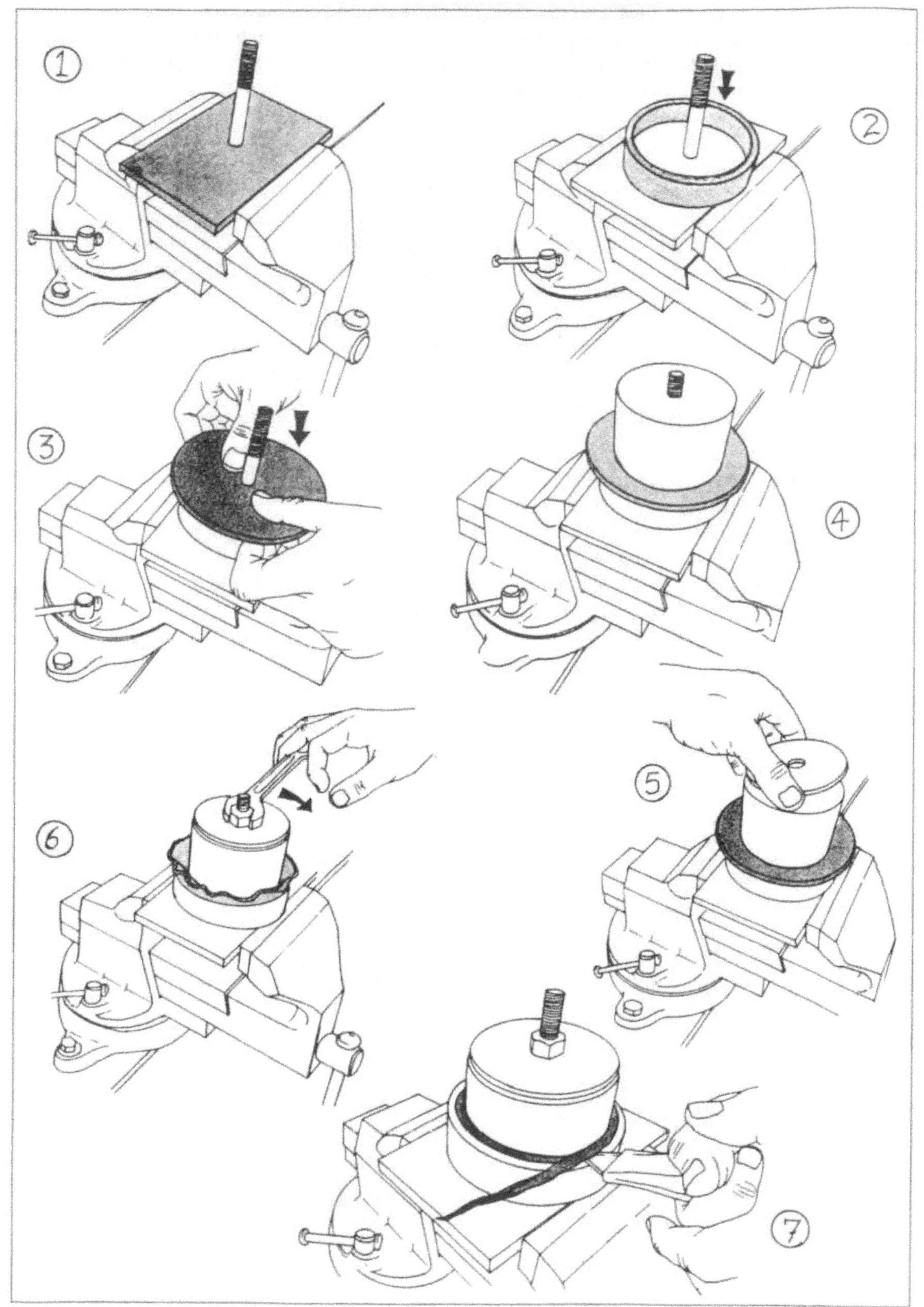


***Figure 28: Forming the leather pump cup***

The tool kit contains two moulds used in forming the leather pump cups. These consist of a wooden male mould, a steel ring (or female mould), a threaded spindle mounted on a base plate, a steel disk, washer and draw-down nut.

Soak the leather disks in water for at least one day, or until they are saturated with water before attempting to mould them. Clamp the base plate in a vice *(Figure 28, Drawing 1)*, and place the female mould ring on the base plate, approximately centred around the spindle *(Figure 28, Drawing 2)*. Place a wet leather disk, smooth side downward, on top of the ring *(Figure 28, Drawing 3)*, and put the wooden male mould on the spindle *(Figure 28, Drawing 4)*, followed by a large flat washer (a piston washer is ideal for this purpose) *(Figure 27, Drawing 5)*, and a nut. Tighten the nut and check that as the leather begins to deform, it does so evenly, all round, and not just one side *(Figure 27, Drawing 6)*. Continue to tighten the nut. If the nut is extremely tight, the leather may be too thick for the wooden mould, and a slightly smaller mould may need to be made to use this leather. On the other hand, if the nut is very loose the leather that has been selected may be too thin to make good quality cups.

Tighten the nut until the wooden mould has entered the ring as far as it will go. Then, use a sharp knife to trim the edge of the cup, level with the top of the steel ring *(Figure 28, Drawing 7)*.

Loosen and remove the nut from the spindle, and pull the wooden mould, leather cup, and steel ring off the spindle as a unit, that is, without removing the ring from the leather cup and wooden mould *(Figure 29, Drawing 1)*. Place the mould in a sunny place for at least one half hour before removing the cup from the mould. Although the cup will dry more quickly when it is released from the mould it must be dry before it is removed in order to maintain its shape. Large-scale pump producers making more than one or two pumps per day should make additional moulds to give the leathers ample time to dry *(Figure 29, Drawing 2)*. To remove the completed pump cup from the mould, reverse the mould, and lightly set the ring on the open jaws of a vice *(Figure 29, Drawing 3)*. Smartly rap downwards on the leather using a wooden dowel, about 1½"–2" in diameter, and the pump cup and male mould will drop out of the female mould.

***Figure 29: Drying the pump cups and releasing from the mould***

The leather cups should be completely dry before they are treated. The best treatment is to rub animal fat (tallow, lard or fish oil) into the leather by hand. The fat may also be heated slightly, and the leather allowed to soak in it. ***Do not use motor oil or any other kind of mineral oil, since these do not protect the leather as well, and may actually contribute to rotting.*** Also, some mineral oils make the pump very hard to operate because they form a sticky film on the cylinders in the presence of water. Some vegetable oils (such as sunflower oil) will also form a sticky film, especially when the pump is not used for a few days.

If the pump cup begins to lose its shape during the drying or treatment process, its shape can be restored by clamping it between two piston washers on the end of a piston rod and inserting it into the mouth of a cylinder. When a pump is stored for any length of time, the pistons should remain in the cylinders to help preserve the shape of the cup.

Paint the piston disks and pump rods with anti-rust primer, and a coat of enamel. Do not paint the leather cups.

The pistons, piston rods and pump cups are then assembled into a completed unit. First, fix the piston rod vertically in a bench vice, and place one of the smaller disks over the threaded end of the piston rod *(Figure 30, Drawing 1)*. Then place a leather pump cup downwards over piston rod, ensuring that it rests flat on the metal disk *(Figure 30, Drawings 2 and 3)*. The larger, intermediate disk is then placed over the piston rod *(Figure 30, Drawing 4)*, and the second pump cup is placed upwards over the piston rod *(Figure 30, Drawing 5)*. The second of the smaller disks is placed on top of the second pump cup *(Figure 30, Drawing 6)*, and a nut threaded over the end of the rod, and tightened against the metal disk *(Figure 30, Drawing 7)*. Use a double 12mm nut made by threading two nuts tightly against each other on a bolt, and weld them together. Remove the nuts form the bolt and fit onto the piston rod. Do not over-tighten the nut, as this may cause the leather to distort. This completes the assembly of the piston unit, which is then removed from the vice.

The operations described in this section require about 2 man-hours per pump, for small series of pumps made by suitably equipped artisans. For large series of pumps, the time may be reduced to approximately $1^1/_2$ hours per pump.

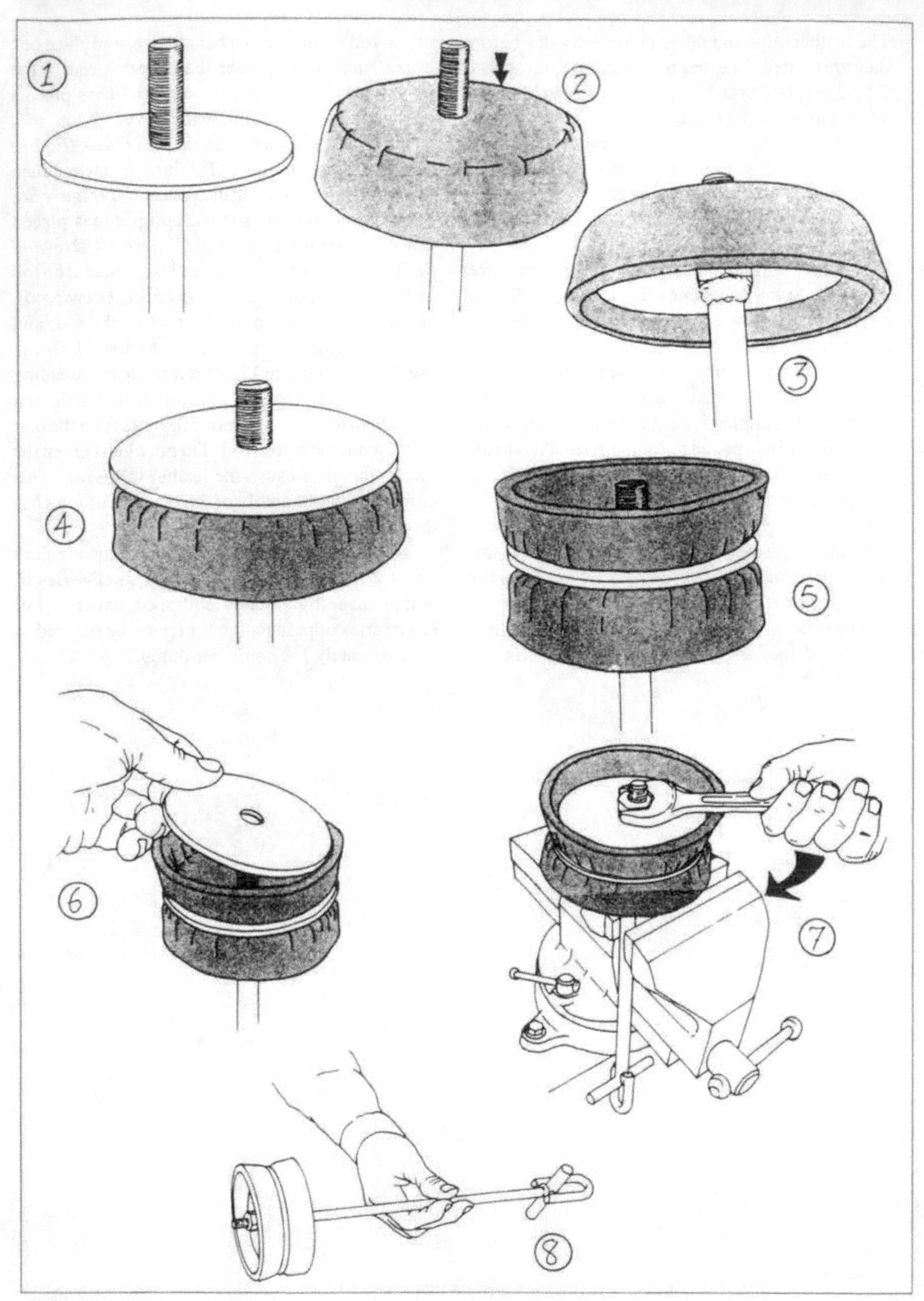


***Figure 30: Assembling the piston unit***

# 5. Making and installing the valves

Four valves are used in the treadle pump: one intake and one outlet valve in each cylinder. Making these valves is simplified because they are identical.

Each valve consists of two parts:

- a rubber valve flap
- a metal clamping bar

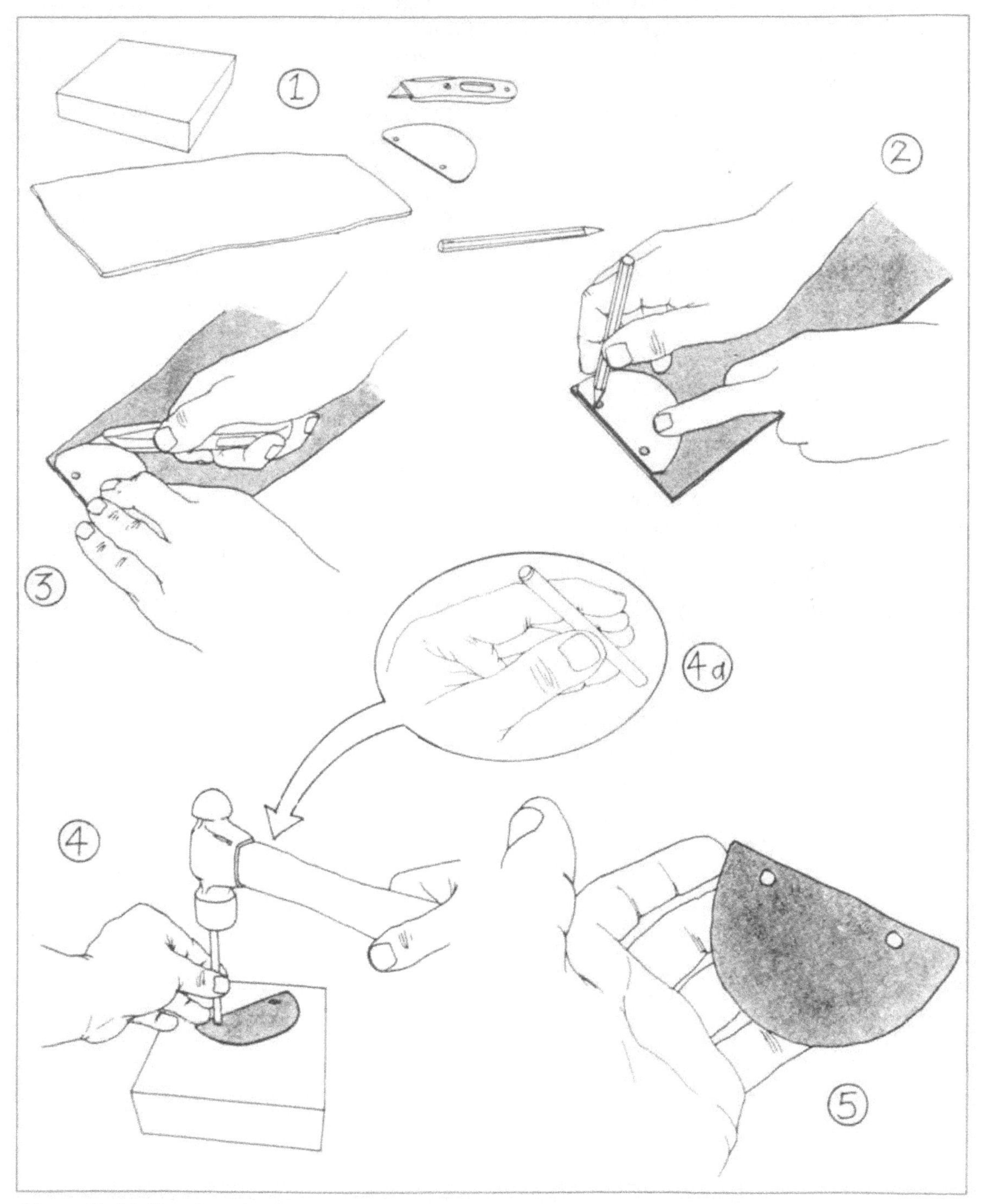


*Figure 31: Making the valve flaps*

Each valve is attached to the pump body with two bolts, which clamp both inlet and outlet valve flaps in each cylinder: thus each cylinder uses only two bolts to attach the two valve flaps.

All of the valves are simple check valves, opening and closing in response to the water pressure acting upon them. The outlet valve opens away from the cylinder and lets water out of the cylinder when the piston is moving downward. The inlet valve opens into the cylinder and lets water into the cylinder when the piston moves upward. The two inlet valves draw water from the inlet chamber of the valve box that connects with the intake pipe. The two outlet valves discharge water into the outlet chamber, connected to the outlet pipe of the pump. Figure 32 illustrates the layout of these valves, and the means by which they are attached to the valve box.

***The rubber valve flaps are cut and punched from 2.0–3.0mm thick truck inner tube.*** Although this may be hard to obtain, do not use inner tube from a car tyre, as this does not have the necessary stiffness to ensure that the outlet valves close easily, and is less durable. Torn and otherwise badly damaged tubes can be used so long as the flaps are cut from undamaged portions that are of uniform thickness. Avoid parts of the inner tube which have seams or ribs moulded into the surface, as this will prevent the valves closing fully.

A half-round template is included in the toolkit *(See template in Figure 31, Drawing 1)*. Placing the template over a strip of inner tube, which is itself resting on a flat wooden block, mark the position of the clamping holes with a ball-point pen *(Figure 31, Drawing 2)*, and cut firmly around the outer edge of the template with a sharp knife *(Figure 31, Drawing 3)*. Using a 4mm diameter hole punch, similar to that used for making the hole in the leather pump cups, tap holes into the valve flaps in the positions already marked in ball point ink *(Figure 31, Drawing 4)*. This completes manufacture of the valve flap *(Figure 31, Drawing 5)*.

Two valves are fitted into each cylinder: the inlet valve is positioned on top of the valve plate, and the outlet valve is positioned below the valve plate *(Figure 32, Drawing 1)*. ***Four clamping bars are made from 10mm (3/8") mild steel square bar, each cut to a length of 80mm, and drilled with two 6.5mm (1/4") holes spaced 48mm apart. A jig for drilling the clamping bars is included in the toolkit. Two 6mm diameter by 40mm long bolts are inserted into each of the two clamping bars,*** and their heads welded to the bars to prevent rotation when being installed in the pump. These clamping bars are installed inside the cylinders, where they will serve both to hold the inlet valves in position, and to stop the piston rods and the leather pump cups striking the valve plate at the bottom of piston travel.

After the welds have cooled, check to see that the bolts are parallel, and may be easily inserted into a plain clamping bar (one with holes drilled at 48mm distance, but without bolts welded in place). It may be necessary to bend the bolts slightly to achieve an easy fit. Remove the plain clamping bar and fit a rubber valve flap over the bolts. Slide the assembly down inside a cylinder, and pass the bolts through the 6.5mm holes in the valve plate. The rubber

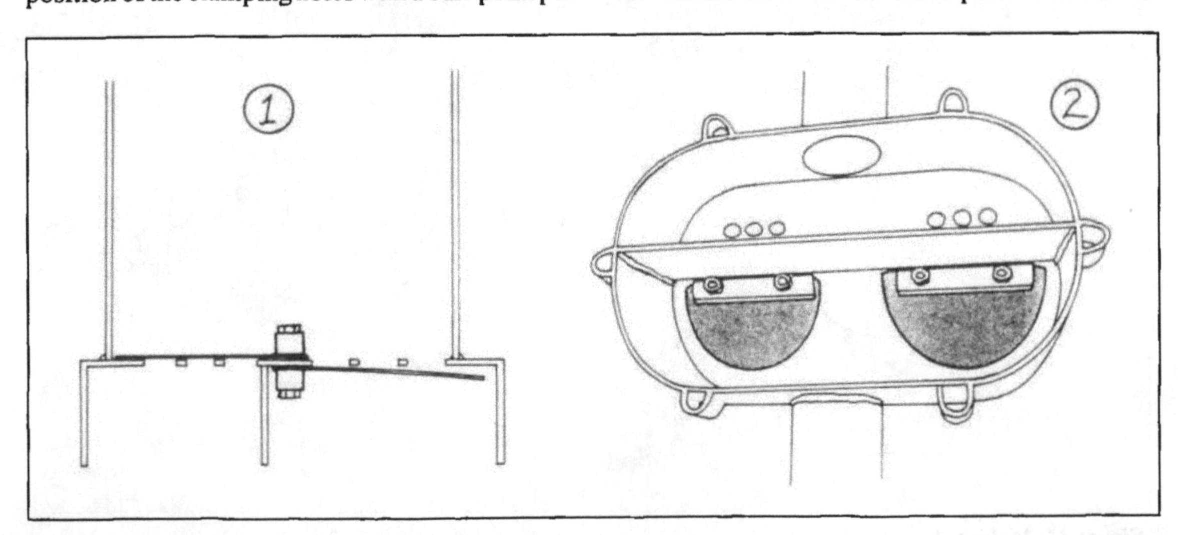


***Figure 32: Installing the valves in the valve box***

valve plate should be installed on the inlet side of the pump, where there is enough room for it to seat without touching the cylinder wall. While holding the inlet valve in position with one hand inside the cylinder, place another valve flap over the outer valve ports under the valve box. Place a plain clamping bar over the protruding bolts and tighten two 6mm nuts lightly on the bolts to hold the flaps in place. Do ***not*** tighten the bolts to the point that the rubber valve flaps begin to distort, as this will prevent the valves closing properly. To ensure that the nuts do not come loose, the end of the bolts should be smeared with contact (rubber) cement. *Figure 32, Drawing 2* shows the discharge valves as they appear from underneath the pump.

Making and installing the valves and the valve clamps generally requires one hour per pump when performed by experienced workers. There are no time savings to be gained between making small and large series.

# 6. Making the wooden parts of the pump

Three parts of the pump are made from wood: the treadles, the pulley, and the baseboard. Wherever possible, all of these parts should be made from good quality hardwood, such as the redwood used for furniture making in West Africa. Hardwood is more expensive than more easily obtained softwood, but it will last much longer against the effects of friction, water and sunlight than softwood. Also, the quantity of wood required is small. In Mali, where hardwood is extremely expensive, being entirely imported from the Ivory Coast, the wood cost is roughly CFA fr 4000 ($8.00) per pump. A wood of medium hardness, such as mahogany, may be used for the treadles and the baseboard, but the pulley ***must*** be made from the hardest wood available. If such wood cannot be procured, the pulley must be made from cast iron or cast aluminium, although this will probably increase the cost of the pump.

***The pulley is made from a small piece of hardwood plank, measuring no less than 40mm thick, 150mm long by 150mm wide.*** If the pulley is to be made on a lathe, drill a 12mm diameter hole through the centre of the piece of wood. Mount the piece of wood on a 12mm nut and bolt held between the chuck and the tailstock centre. If the pulley is to be turned on a wood lathe it can be glued to the face plate with a piece of parting paper, as would be done in turning a wooden bowl. Alternatively, it may be secured between the tailstock and the face plate *(Figure 33, Drawing 1)*.

Turn the pulley to a diameter of 140mm and cut a groove in its rim to a depth of 12–15mm *(Figure 33, Drawing 2)*. Ensure that the groove is at least 15mm wide, and has a rounded section, to minimize rope friction. It is helpful if the cutting tool has a rounded nose to facilitate cutting the groove to a curved section. Remove the pulley from the lathe, and, while holding it in a clamp, drill the centre hole to the same diameter as the pulley shaft which is mounted on the handle *(Figure 33, Drawing 3)*. This should not be less than 18mm. Do not attempt to drill this hole with an electric hand drill or brace, because it is difficult to ensure that the hole is bored perpendicular to the face of the pulley. If the hole is at an angle, the pulley will wobble, and may cause the rope to dismount. Once the pulley is completed, immerse it in a bath of used oil, and put

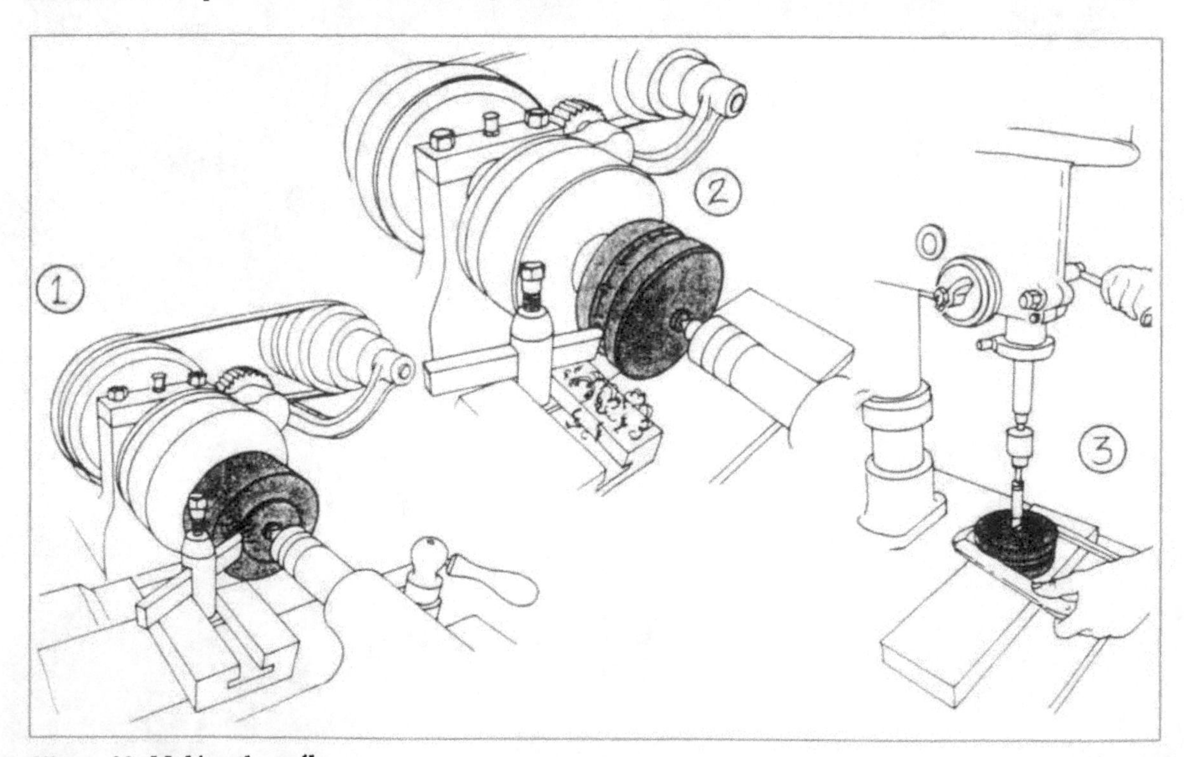


*Figure 33: Making the pulley*

weights on top of the pulley to ensure that it remains immersed for at least a week. This will permit the oil to penetrate the wood, and act as a long-life lubricant. It will also ensure that any oil which is subsequently added to lubricate the pulley is not immediately absorbed into the wood.

***The treadles can be cut by sawing a 40mm thick hardwood plank into pieces 80mm wide, or by sawing an 80mm square hardwood post into two halves, roughly measuring 40 by 80mm. If a softwood must be used for the treadles, increase the thickness from 40mm to 50mm to compensate for its reduced strength. The treadles should be 1 600mm long.***

The treadles are first bored edgewise with a 12mm drill at a distance of 50mm from one end *(Figure 34, Drawing 1)*. One of the treadles is then fitted over the T piece on the treadle support, and positioned parallel to the ground, resting above the cylinder (Figure *34, Drawing 4)*. The centre position of the cylinder axis is marked on the treadle, and the treadle is removed from the treadle support. Two 10mm diameter holes are drilled, 50mm apart, and with centres that are 25mm on either side of the marking which indicates the centre of the cylinder axis, along the longitudinal axis of the treadle *(Figure 34, Drawing 2)*. The wood between these holes is removed with a 10mm chisel, making a slot through the treadle. This is sufficient to allow the 8mm thick loops which form the top of the piston rods to pass easily through the treadle. The end of the slot should be no closer to the end of the treadle than 350mm *(Figure 34, Drawing 4)*.

If many of these treadles are to be made at once, these slots and the treadle pivot holes can be conveniently and accurately made on the mortising table of a multi-purpose woodworking machine. If a hand-held drill or brace is used to make these holes, be careful to verify that the drill enters the wood perpendicular to its surface.

The last wooden part to be made is the baseboard, onto which the pump is mounted, and which

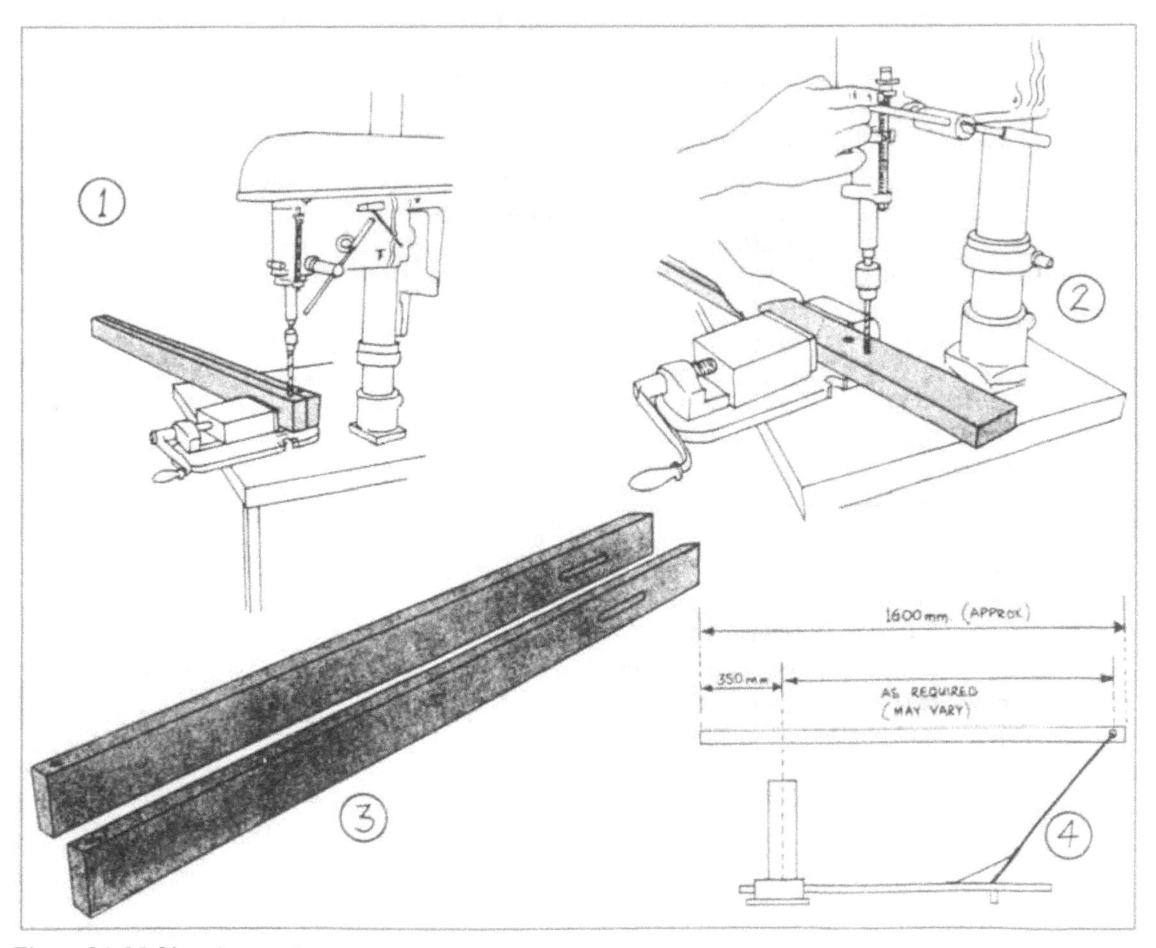


***Figure 34: Making the wooden treadles***

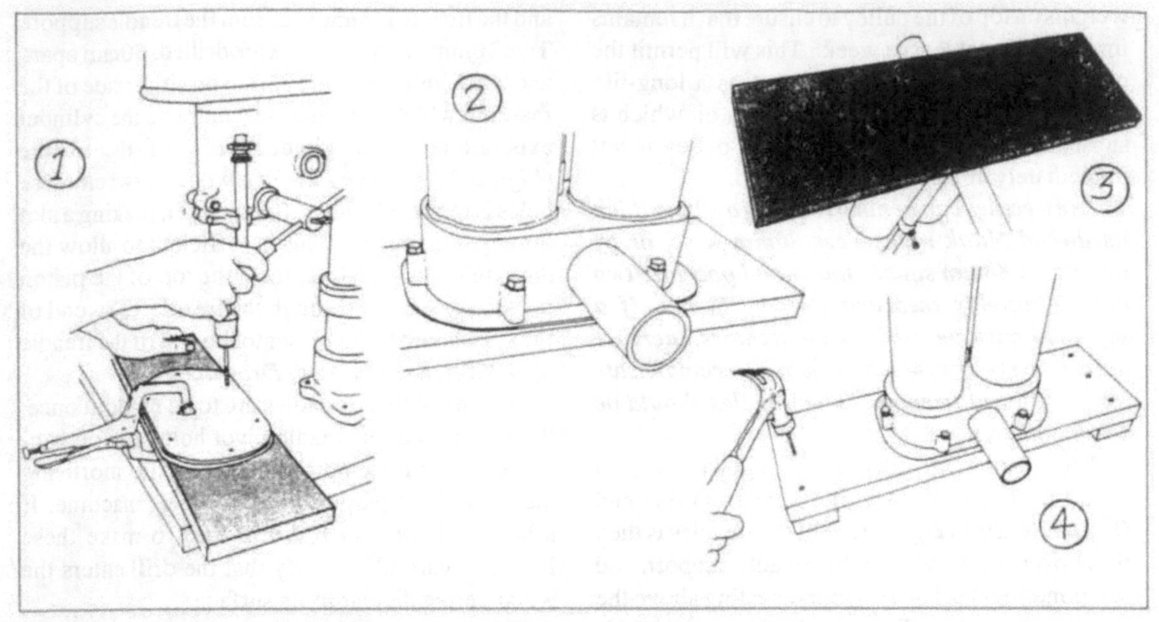


*Figure 35: Making an installing the baseboard on the pump body*

forms the bottom of the valve box.

***In making the baseboard, select a plank roughly 200mm wide and 30mm thick, that is thoroughly dry and flat.*** If the only plank that is available is warped or curved, it should be planed flat on both sides before use. Planing will reduce the thickness of the board, and it should not, under any circumstances be less thick than 25mm after planing.

The toolkit includes a template for drilling the six holes in the baseboard that correspond to the six 8mm diameter studs on the pump body. It is important to use the template when drilling these holes, because the studs should slide easily into the baseboard holes, allowing it to be easily removed. Clamp the baseboard to the template *(Figure 35, Drawing 1)*, and sandwich a piece of heavy duty rubber coated canvas or nylon hose between the template and the baseboard.[21] This permits the holes in the baseboard and the sealing gasket to be drilled at the same time, and in exactly the same position. Use a drill press to bore these holes to 8.5mm diameter. If a drill press is not available, the template may be modified by welding guide bushings to it.

Two cleats, or feet, made from 40mm square hardwood are then nailed or screwed to the baseboard *(Figure 35, Drawing 4)*. The cleats reinforce the baseboard, increase the stability of the pump, and help to keep the baseboard and the valve box studs clean and dry when the pump is in use. If the wood used in making the baseboard is prone to split, or is very hard, it should be pre-drilled before nailing or screwing the cleats to the baseboard. The durability of the baseboard, as well as other wooden parts of the pump can be increased by painting them with several coats of varnish, which will also enhance the appearance of the pump.

The operations described in this section require 1½ man-hours labour per pump, when pumps are made in series by experienced workers. This can be reduced to under one hour if the pumps are made in large series with appropriate mechanization, particularly a multi-purpose woodworking machine.

---

[21] Various materials, including heavy leather and inner-tube rubber, may be used to make the gasket, but rubber-coated canvas or nylon hose seems to work best. This nylon hose is often sold for use with petrol or diesel driven irrigation pumps, and may be very expensive. However, a 280mm length of 75mm diameter hose is sufficient to make one gasket.

# 7. Final assembly of the pump

When the baseboard is completed, six bolts measuring 8mm by 70mm are inserted into the six pockets, or bushings, on the walls of the valve box. The rubber gasket is fitted over these bolts, through the holes drilled in the previous step, and the baseboard offered up to allow the bolts to pass through from top to bottom. Washers are placed over the bolts, and nuts threaded on, and tightened down. The nuts should be tightened from opposite sides of the valve box, to ensure that there is no distortion in the valve box, or uneven pressure exerted on the gasket *(Figure 35, Drawings 2 and 3)*. Because the intake side of the pump works at less than atmospheric pressure, the gasket will tend to bulge up into the valve box on this side. While this is not likely to affect the operation or efficiency of the pump, it can be prevented by cementing the gasket to the board with contact (rubber) cement. Be careful accurately to line up the holes in the gasket with those in the baseboard when gluing the two together.

Once the pump is mounted on the baseboard, set the pump in its upright position, and insert the handle pipe. Lubricate the pulley shaft with mineral or animal grease and slide the pulley on to the shaft, followed by a large flat washer *(Figure 36, Drawing 1)*, and cotter pin *(Figure 36, Drawing 2)*.

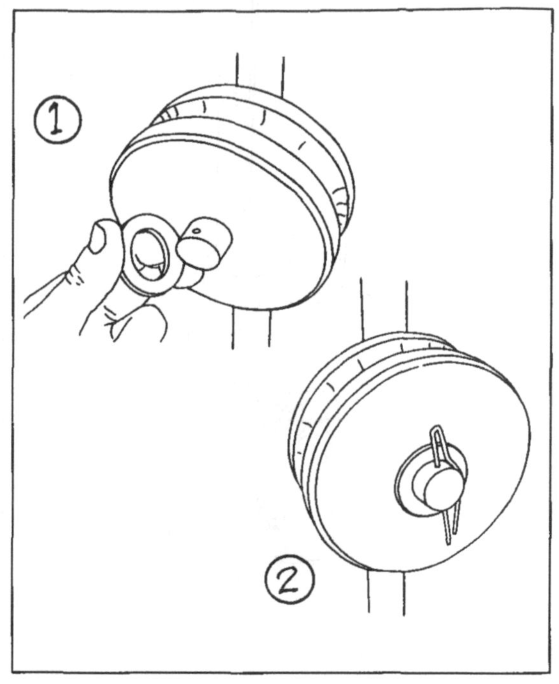


*Figure 36: Installing the pulley on to the pulley shaft*

Next, assemble the piston assemblies, and insert these into the cylinders. Care should be taken to ensure that the lower pump cup is gently manipu-

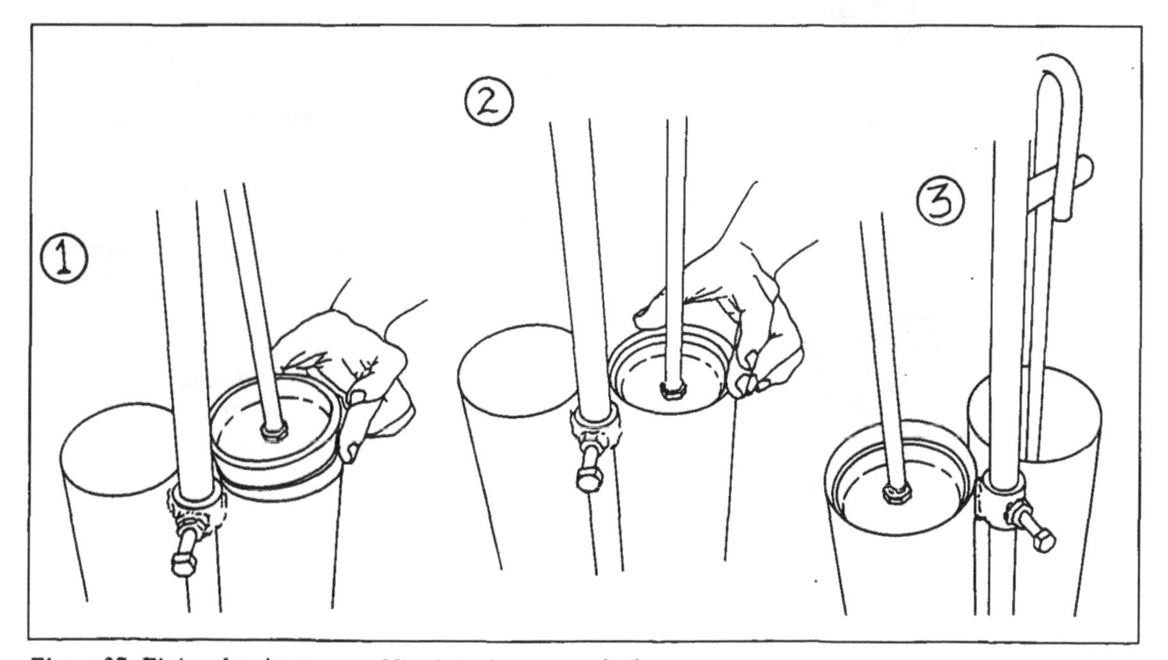


*Figure 37: Fitting the piston assemblies into the pump cylinders*

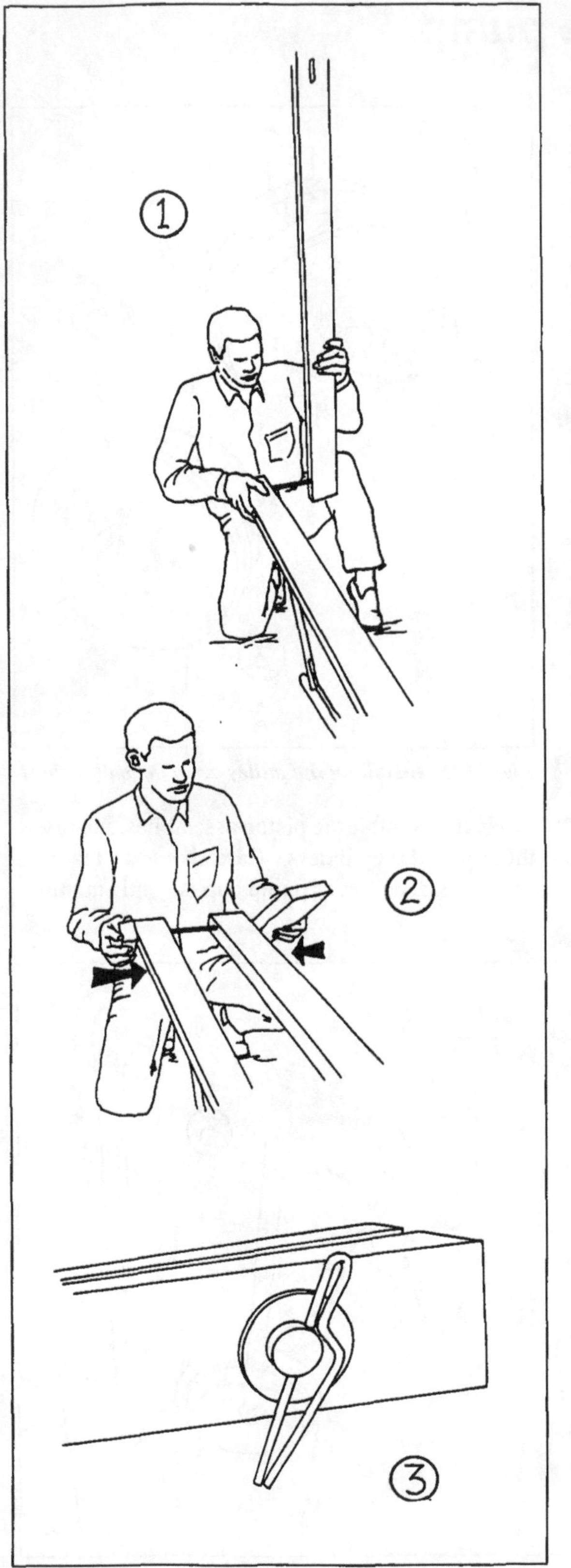


***Figure 38: Fitting the treadles on to the treadle support***

lated into the cylinder to avoid damage to the softened leather (*Figure 37, Drawing 1*).

Next, slide the two treadles on to the treadle pivot shaft *(Figure 38, Drawing 1)*. If the treadles fit tightly on the shaft, it may be necessary to drive the treadles on to the shaft by tapping them with a mallet, or block of wood *(Figure 38, Drawing 2)*.

Make sure that the ends of the treadle pivot shaft are chamfered, to facilitate the fitting of the treadles, and that the treadles are properly aligned before tapping them on to the shaft. It may be necessary to bend the treadle pivot shaft, either before or after fitting the treadles, so that the treadles find their correct position over the cylinders. The treadles can rub lightly against the handle pipe, but should not be more than two centimetres from it. If this is the case, remove the handle pipe from the pump, raise the treadle to be adjusted, and push it towards the middle of the pump until its position is correct. Then, re-install the handle pipe and tighten its attachment bolt. Install cotter pins (made from 4mm round bar) to hold the treadle on the pivot shaft *(Figure 38, Drawing 3)*.

The final step in assembling the pump is to install the pulley rope that connects the two pistons and treadles. First of all lower one piston until it touches the bottom of its cylinder, and raise the other until it reaches the top of its cylinder *(Figure 39, Drawing 1)*. ***Using polypropylene rope of between 10–12mm diameter***, tie one end of the rope to one of the piston rods with a double knot, securing the end of the rope by means of wire cleats *(Figure 39, Drawing 2)*. Pass the rope over the pulley and through the other piston rod, taking care to lift the raised piston cups slightly out of the cylinder, while making sure that the other piston remains at the bottom of its stroke *(Figure 39, Drawing 3)*. This ensures that when the pump is in operation the pistons are unlikely to bottom on the valve plate, since to do so would raise the operator's leg to an excessive height on the upstroke. Tie the rope as for the first piston. Now the pump is ready for use *(Figure 39, Drawing 4)*.

The final assembly of the pump should require between ½ hour and 1 hour.

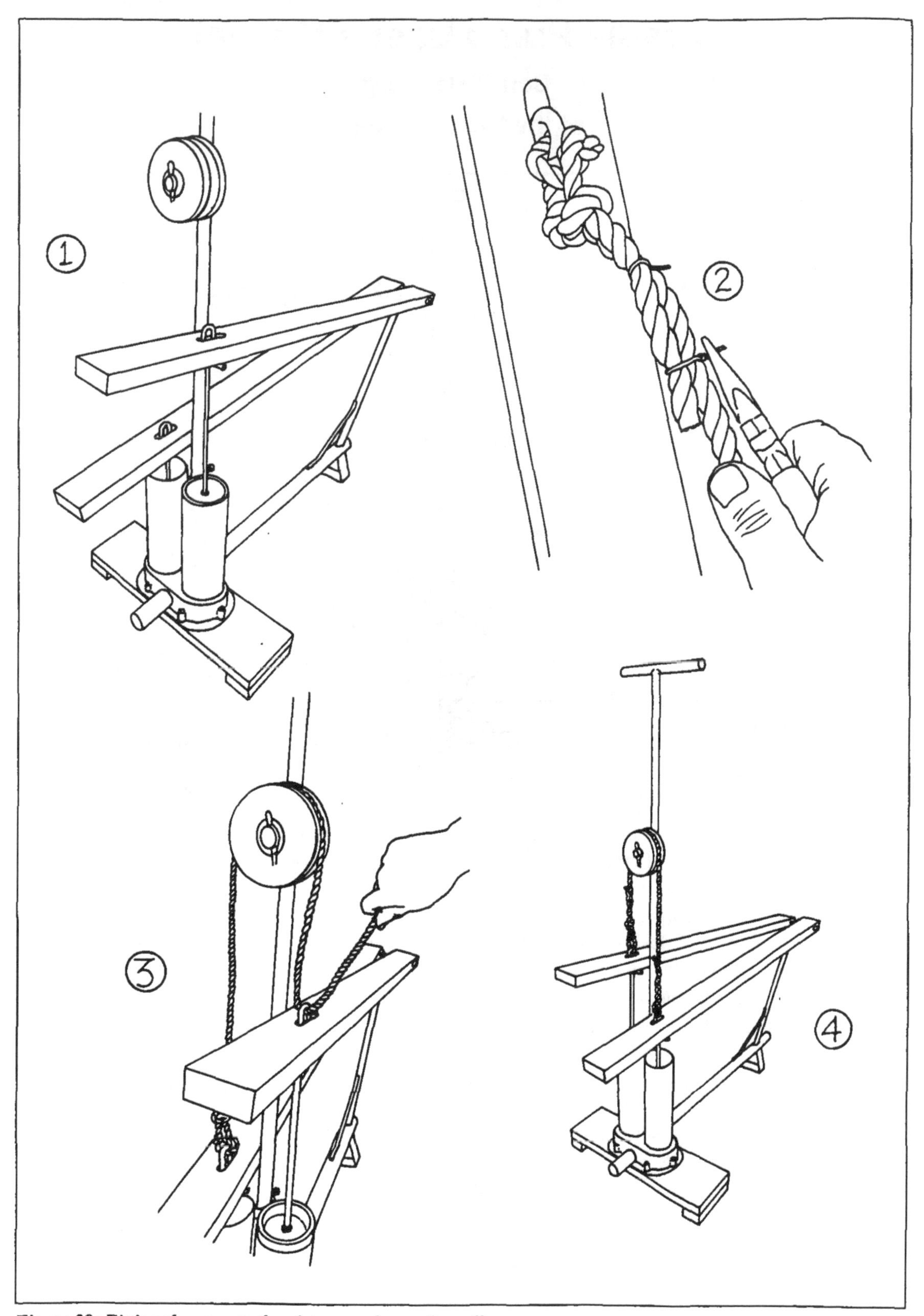


*Figure 39: Fitting the rope to the pistons and over the pulley*

# Treadle Pump User's Manual

Carl Bielenberg

## I. General characteristics of the treadle pump

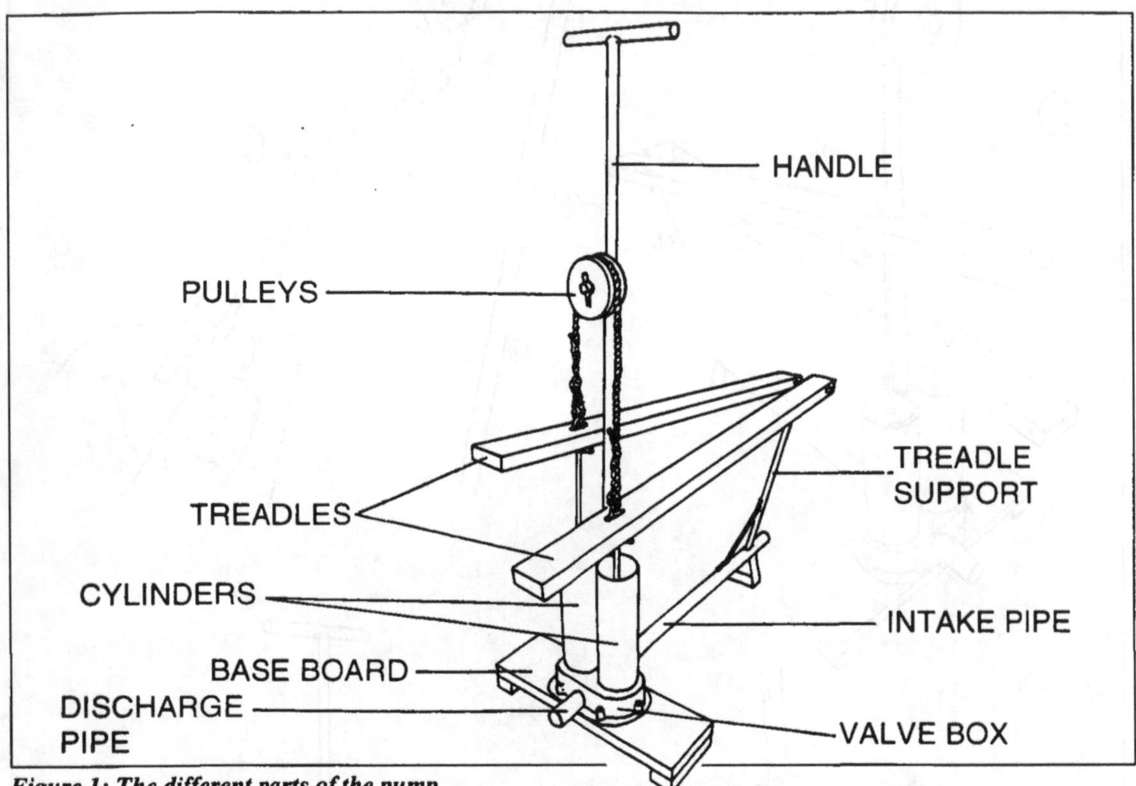


*Figure 1: The different parts of the pump*

The treadle pump is designed to help market gardeners irrigate small areas of crops. The pump is human powered, to eliminate the fuel and oil costs associated with motorized pumps, and reduce the crop damage and income losses associated with mechanical breakdowns and interruption of the fuel supply. The treadle pump is simple and efficient. It can pump 5 000–7 000 litres (or 5–7 m$^3$) of water per hour. If the pump is properly adjusted and maintained, the pump can be operated several hours a day without tiring the workers. The leg muscles, which are the most powerful of the human body, provide the power to drive the pump. In addition, the pump is operated in a natural way, as if walking. With four hours of pumping per day, depending on the soil permeability and the water needs of the particular crop, the pump can irrigate up to 0.5 hectares of land in the dry season.

The treadle pump can raise water by suction and by pressure, just like a motorized centrifugal pump. It can raise water from a well when the water is less than seven metres below ground level. It can also pump water through a distribution pipe. The pump can be used by one or two adults or up to four children. The total vertical distance that water can be pumped is limited by the total weight of the pump operators. Two adults weighing a total of 100kg can pump water up to 10 metres vertically. One adult weighing 50kg can lift water only five metres. If the pipe used with the pump is too small or the pump is not maintained and adjusted properly, the pump will not be able to lift water this high and the work is likely to be tiring. Figure 1 shows different parts of the pump. Figures 2a and 2b show the treadle pump installed next to a well, discharging water into an basin, and being operated by one or two workers.

## 2. Preparation and installation of the Treadle Pump

*Figure 2 a*

*Figure 2 b*

***Figure 2: The Treadle Pump installed in the field***

Before using the pump, the user must grease the pulley shaft with good lubricant such as automotive grease, lard, cooking fat or shea nut butter. Remove the pulley to be sure that the shaft and cotter pin are well lubricated, and replace the pulley with its washer and cotter pin *(Figure 3, Drawings 1 and 2)*.

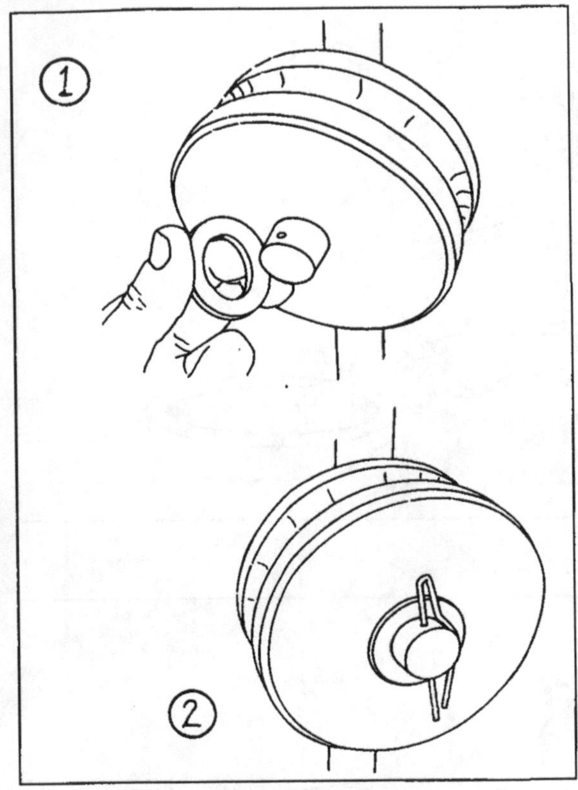


***Figure 3: Pulley removal and installation***

Next, the user should remove the two pistons from their cylinders to check that the valves work properly. The outlet valve in each cylinder, on the same side of the pump as the outlet pipe, should open when pressed down with a finger, through any one of the 12 port holes, and should re-close automatically, to within a very close (2–3mm) distance to the valve plate. The inlet valve rubber on the same side of the pump as the intake pipe should open easily when lifted upward. Sometimes, these valves may stick closed on a new pump if the valves are installed before the paint has entirely dried. If this or any other problem with the valves is noticed, refer to the section of this manual on maintenance and repair of the pump (Section 5).

The pump is connected to the source of water by a suction pipe. The suction pipe must be rigid, or of the reinforced flexible type so that it will not collapse. In some cases, a delivery pipe carries water to the farm or canal, away from the pump. The delivery pipe may be flexible, but should not collapse when bent, or it will restrict the flow of water, or make pumping difficult. Rigid PVC pipe with an inside diameter of at least 45mm is recommended for use with the treadle pump because it is relatively inexpensive, lightweight, and enables the water to flow very freely. Figure 4 shows how to connect lengths of pipe in a way that is airtight and easily detachable:

- one end of the pipe must be heated carefully and rapidly over an open flame, which can be created using waste paper. The PVC will rapidly become soft and flexible.
- when this happens slip the stiff end of an ***unheated*** length of PVC pipe into the open end of the heated pipe, for a distance of about 50mm.
- wait for the pipe to cool, at which time it will regain its stiffness.
- wrap strips of inner tube around the joint, making sure that they are tightly stretched.

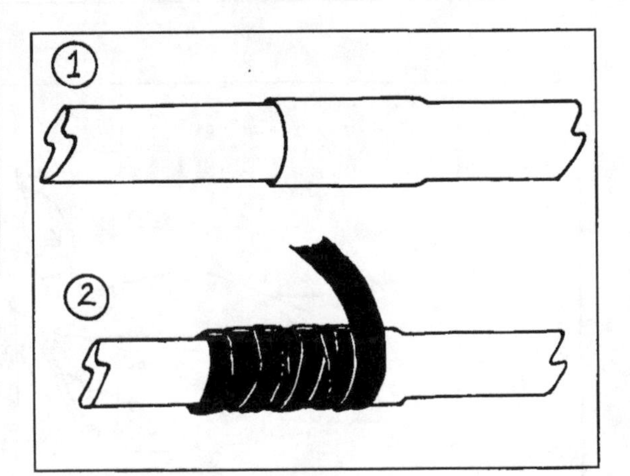


***Figure 4: Joining two lengths of PVC pipe***

To prevent stones and other objects from entering the pump it is important to install a fine metal screen (1mm mesh) at the intake of the suction pipe. The intake screen must be supported above the bottom of the river or well to prevent it becoming clogged with mud. If the pump is installed in a well, an intake screen is not necessary, so long as the lower end of the pipe is suspended at least 200mm from the well floor.

Each piston has two leather cups, one turned toward the bottom of the cylinder, and the other turned toward the top. The first cup maintains a seal

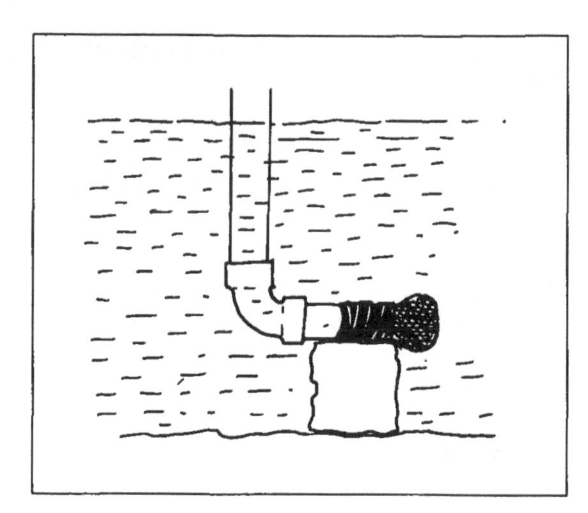

*Figure 5: Intake screen and support*

under pressure when the piston goes down. The second cup maintains a seal under suction when the piston goes up. If the pump is new, or has not been used for several days, the leather piston cups will dry and will not provide a good seal. Just before using the pump, remove the pistons from their cylinders and put them in a bucket of water. The pistons can be removed from the pump without detaching the rope which connects them together. If, after 15 minutes soaking, the leather cups are still stiff, work them gently by hand as shown in Figure 6. As the leather cups become soft, they fit the form of the cylinders well, and provide a good seal. When replacing the pistons in the pump cylinders, it is important to take care that the lower cup is not bent back or damaged as it is inserted in the cylinder mouth.

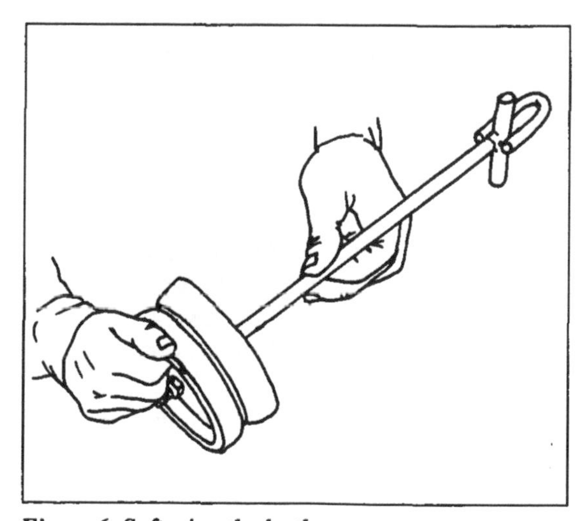

*Figure 6: Softening the leather pump cups*

## 3. Starting the pump

The proper way to start the pump is easily learned. When one begins operating the pump, the treadles move very freely, since there is no water in the pump or suction pipe. Starting the pump, the operator must work the treadles very quickly to evacuate the air and pull the water up. To facilitate the sealing of the pistons against the entry of air, many treadle pump users use a watering can to fill up the cylinders from the top, above the pistons. This is only necessary at the time of start-up: when the cylinders are full of water they tend to create their own seal. If constant topping up remains necessary when the pump is working, the seal of the pump cup to the piston wall is not good enough. The rapid treadling action is only necessary for a few moments: the amount of time it takes the water to reach the pump cylinders. As the water fills the pump, the movement of the treadles will become less free, and the operator can establish a comfortable rhythm. The operator should not hit the pistons against the bottom of the pump, to avoid damaging the valve box. The best way to pump is a comfortable motion that provides a constant stream of water. The users should try to move the pistons through their full stroke without hitting the bottom of the cylinders.

If the pump does not start at all or does not work well, check that there are no leaks in the suction pipe, and that there is a good seal between the pump and its baseboard, and between the leather cups and the cylinders. A sound of air whistling when treadling stops indicates a leak in the suction pipe or baseboard seal. A leak in the seal can usually be stopped by lightly tightening one or more of the nuts located underneath the baseboard, on the side of the pump where the leak exists. Leaks at the leather cups can be found if, during pumping, there is a spray of air and water there, or if the upper leather cups are not touching the cylinders all round. Leaks around the leather cups can usually be eliminated by removing the pistons, and expanding the leathers by working them with the fingers as illustrated in Figure 6.

If water reaches the pump, but the pumping seems difficult, and the treadles move slowly, check the intake to the suction pipe. If the suction intake does not sit on the bottom of the well or river, and the intake screen is not clogged, the leather cups may require lubrication. Before lubricating the leathers, read the section of this manual on maintenance and repair.

## 4. Different methods of irrigation

The treadle pump can be used for several methods of irrigation. On clay soils, water can be distributed from the pump in hand-dug canals passing between raised beds or in depressed beds that are periodically submerged in water. On sandy soils, a pipe is often used between the pump and the crop, to avoid heavy losses from water soaking into the soil.

Although farmers with motorized pumps commonly use a long flexible hose and pinch the end of the pipe to create a spray, this methods requires a lot of pressure and would be tiring for treadle pump operators. Distributing water by canals requires the least pressure and the smallest amount of pipe and is therefore recommended wherever the soil has enough clay to make this possible. If tubing has to be used between the pump and the field it should consist of rigid PVC plastic or metal pipe with an inside diameter of 45mm or more. In addition, the field and the source of water should be planned so that the length of pipe is as short as possible. If the field has a gradual slope, the well should be located at the highest point in the field, if possible, so that the water can be distributed by gravity in open canals. If the land is level, the well should be located in the middle of the field.

A method widely used in West Africa, where sandy soils are common, is for the farmer to build a water tank, usually not less than $1m^3$, as close as possible to the centre of the field, and pump water under pressure from the well (which may be a long distance away) to this basin. Water is then distributed to the crop by means of watering cans. Farmers in Togo will often have five or six of these tanks, connected by underground piping, and fed by a single pump located at the centre of the field.

The maximum length of tubing that can readily be used with the treadle pump is about 150 metres, when the land is level. If the field is too far from an existing well, a new well should be dug at the centre or the top of the field. If the water table is found at seven metres or less only at the lowest part of the farm, the pump can deliver water from the well to basins located higher up by pumping to these basins through PVC pipe.

## 5. Maintenance and repair of the Treadle Pump

The parts of the pump that require the most maintenance are easy to see and replace. These are:

- the pulley
- the rope
- the leather pump cups

The pulley shaft should be lubricated every day that the pump is used. The condition of the rope should be checked to avoid the risk that it snaps when carrying the weight of the operators, and, if natural fibre rope is used, checked to see that its length permits each piston to move from the top of its cylinder to the bottom. If the rope has to be replaced, select a strong type that does not stretch. Polypropylene rope with a 10mm diameter – the brightly coloured kind found in many markets, is a very good choice.

The parts that are most likely to cause a malfunction are the leather pump cups. If the pump is used only once or twice per week, the leather cups will dry out quickly. To prevent the leather from drying out, remove the pistons from the cylinders, and keep the pistons submerged in a bucket of water between uses. This method will also allow the inside of the pump to dry out, so it will not rust as quickly. From time to time, check that the leathers still fit properly inside the cylinders. If the leather cups frequently change shape after soaking, and do not fit the cylinders well, you may choose to leave the pistons in the pump. However, the leather cups may require soaking and massaging (as shown in Figure 6) the next time that the pump is used.

If the pumping becomes difficult, check that the intake (suction) pipe and filter are not blocked. If you are using a flexible plastic or rubber hose, check that it is not sharply bent, since that will restrict the flow of water. If you are using inner tube to join lengths of pipe on the intake side of the pump, the rubber may get sucked in between the pipes and restrict the flow. If the pipes are not made of PVC, and do not fit inside one another, short lengths of reinforced plastic or rubber hose should be used to join them. Also, make sure that the filter is properly attached to the intake pipe, and that the lower end of this pipe is suspended above the bottom of the well or river.

If a filter has not been used or has become detached, small stones,[22] gravel or sand may enter the pump, preventing the valves closing properly, and making pumping difficult. These can usually be removed by taking off the inlet and outlet pipes, removing the pistons, pouring water into the cylinders and opening the valves with the fingers. If stones, sand or gravel remains inside the pump, the baseboard must be removed by taking off the six nuts that attach it to the pump.

If pumping is still difficult after taking these steps, the pump leathers probably require lubrication. Shea butter, beef tallow, or rendered butter (ghee) are good lubricants for the pump cups.[23] If you do not need to use the pump for a while, remove the pistons, and allow the leather cups to dry completely, and apply a lubricant. Because the leather is dry, it will absorb more of the lubricant, and the lubrication will last longer. Working the lubricant into the leather, as shown in Figure 6, is beneficial. If you have to use the pump immediately, you can begin pumping after applying the lubricant to the wet pump leathers.

If the treadles move easily, but the pump doesn't lift any water, or only a small amount, then there is probably an air leak at the joints of the suction pipe, the baseboard gasket, or the leather pump cups. A very small leak can prevent the pump from lifting water, especially if the vertical distance from the pump to the water is more than five metres. If the vertical distance to the water is more than eight metres, the treadle pump cannot easily be used, even if there are no leaks.

If the distance is less than eight metres, and no leaks are found, the problem may be caused by the valves. In that case, remove the pistons from their cylinders, and use your fingers to check that the valves are working properly, especially on the intake side, They should open freely and fall by their own weight to the closed position. The two outlet valves should open when pressed with a finger, and spring upwards to within 3mm of the closed position when the finger is removed. If the outlet valves do not close, the pump will be difficult to prime, but may work well when full of water. To prime a pump that has this problem, install a short length of pipe on the outlet side of the pump, tilt it upwards, fill it with water, and then start the pump. Once the pump is started, additional lengths of pipe can be connected as long as water remains in the first pipe. If there is a discharge pipe, the pump can also be primed by removing one piston and pouring water into the cylinder. If some gravel or sand is found inside the pump when checking the condition of the valves, remove the sand or gravel, and be sure that the screen is properly installed.

By removing the base board, or plank, the four valves can be inspected, and easily removed. First, remove the six nuts and washers from the bolts that hold the plank to the pump body, and separate the plank from the pump by tapping it lightly from above with another piece of wood, while supporting the pump off the ground. The two valves that are most visible when looking from inside the valve box are the outlet valves. To remove all four valves, undo the four retaining nuts that secure the two metal clamps to the valve rubber *(Figure 7)*.

Next, lift the clamps from the bolts, and remove the valve rubbers. Push the bolts through the valve plate, and withdraw the intake valves from inside the cylinders. Inspect the valve rubbers. If there are signs of wear, especially around the clamps, or if the rubber has developed a distorted and tired appearance (especially if the valve port holes have started to cut into the valve rubber), they should be

[22] Cases have been recorded of the valve box being completely filled with stones up to 30mm in diameter.

[23] Do ***not*** use kerosene nor waste motor oil to lubricate the leather cups because it will make them wear out quickly.

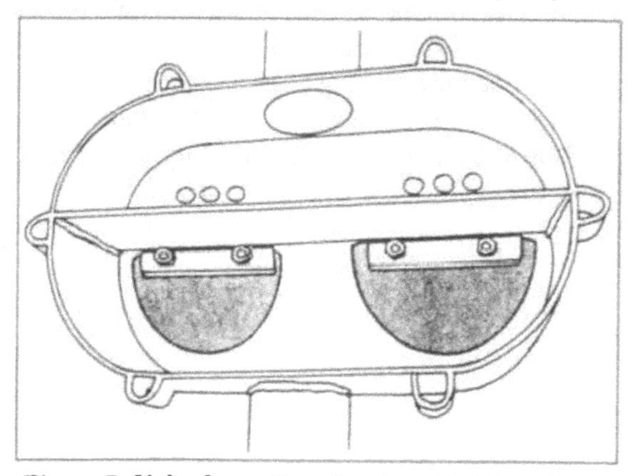

***Figure 7: Valve box with valves installed viewed from below***

replaced. Cut new valve rubbers from truck inner tube ***which is at least 2mm (and preferably 3mm) thick.*** Punch mounting holes in the valve rubbers in positions that correspond to the holes in the retaining clamps, and re-install the valves. Pay particular attention to the appearance of the valve rubbers when they are re-installed. Because they are cut from inner tube, which is curved, they will have a tendency to curve in a particular direction. Make sure that when they are clamped in place that if they curve at all, they curve over the valve port holes, and not away from the valve plate. To install the valves correctly it is best to remove one set of valves from one cylinder at a time. Reassemble the valves so that they are configured in the same way as the other cylinder.

The pump cups should have a thickness of between 4–5mm. The original leather cups are made from industrially treated leather formed in special moulds and then dried and treated with animal or vegetable fat. They are the best leather for your pump, and are available from the pump maker. The leather pump cups can be made by a leather-working artisan or the pump owner, ***if the right kind of leather is available.*** It is impossible to know whether a particular kind of leather produced by a traditional process will work without trying it. Some leathers become very hard the first time they dry out, after having been used for a time in the pump. Other leathers become extremely soft and floppy after a short time in use. The experience of pump makers in West Africa is that leather produced by traditional methods is usually not suitable, and lasts for only a short time. If at all possible, try to obtain leather from a shoemaker, used for making the upper parts of shoes, or from a saddle maker. This leather should be a light brown colour, and should be quite flexible. Soft leather of the type used for making bags, and the hard leather used for making shoe soles are ***not*** suitable.

If the leather cups on your pump are not worn out or cracked, but no longer have the proper form, you can easily remould them yourself. To do this, unscrew the 12mm nut on the bottom of each pump rod that holds the leather cups, remove the cups, and soak them in water for several hours. Then, re-install only one leather between the metal piston disks on each pump rod with the rim of the leather facing upward (toward the treadle end of the pump rod). Tighten the nut to hold the leather securely in position, and push the pump rod into one of the pump cylinders. When the leather cup has entered the cylinder, so that its edge is at the same height as the cylinder rim, use a pair of pliers to press out any bumps or kinks that may have formed in the cup against the side of the cylinder. When the cup begins to hold the proper shape, let it dry for an hour or more before removing it from the cylinder, and allowing it to dry completely in the sun. When fully dry (after at least a day), the cup should be soaked in beef tallow, lard or ghee. It should then be usable. This same process can be used to make new leathers from disks of leather if the proper spares are not available.

If the pump cups continue to give problems, and their edges fall away from the cylinder walls in a series of kinks, this can be remedied permanently by making three washers from truck inner tube that correspond in size to the metal washers that fit inside the pump cups. These disks are fitted inside the pump cups, and the metal washer fitted on top. This raises the metal washer upwards, and helps to support the pump cup walls higher up than normal, pressing them against the cylinder walls. Ensure that you do not over-tighten the retaining nut, because this will cause the inner-tube washers to expand, and press the leather too hard against the cylinder walls.

Any user complaints with the pump should be addressed to the supplier, who will assist you with repairs or, if necessary, help you obtain a refund from the manufacturer, under the terms of your guarantee.

# Appendix A: Engineering Drawing Set

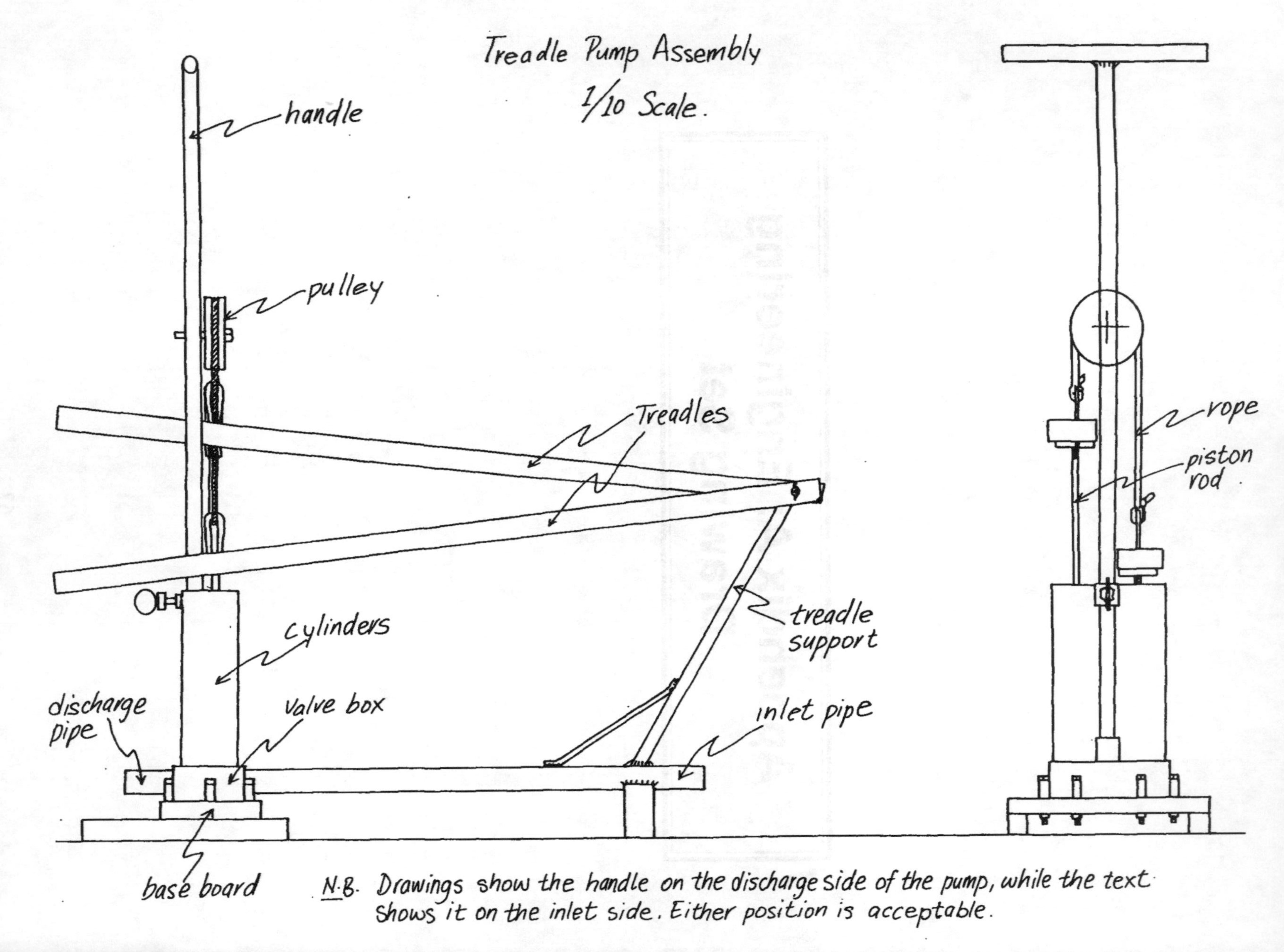
Treadle Pump Assembly
1/10 Scale.
handle
pulley
Treadles
rope
piston rod
cylinders
treadle support
discharge pipe
valve box
inlet pipe
base board
N.B. Drawings show the handle on the discharge side of the pump, while the text shows it on the inlet side. Either position is acceptable.

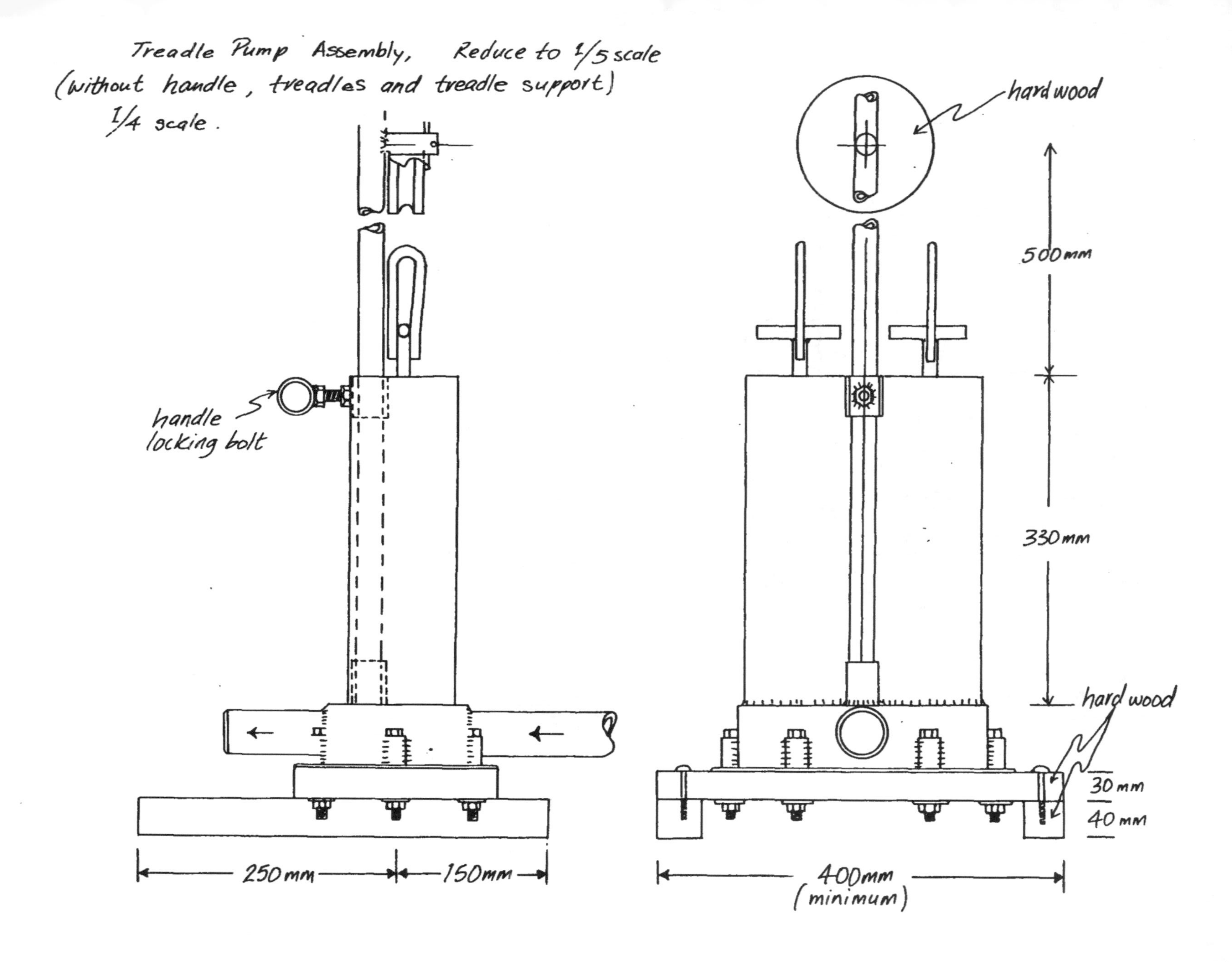
Treadle Pump Assembly, Reduce to 1/5 scale
(without handle, treadles and treadle support)
1/4 scale.
handle locking bolt
250mm
150mm
hardwood
500mm
330mm
hardwood
30mm
40mm
400mm
(minimum)

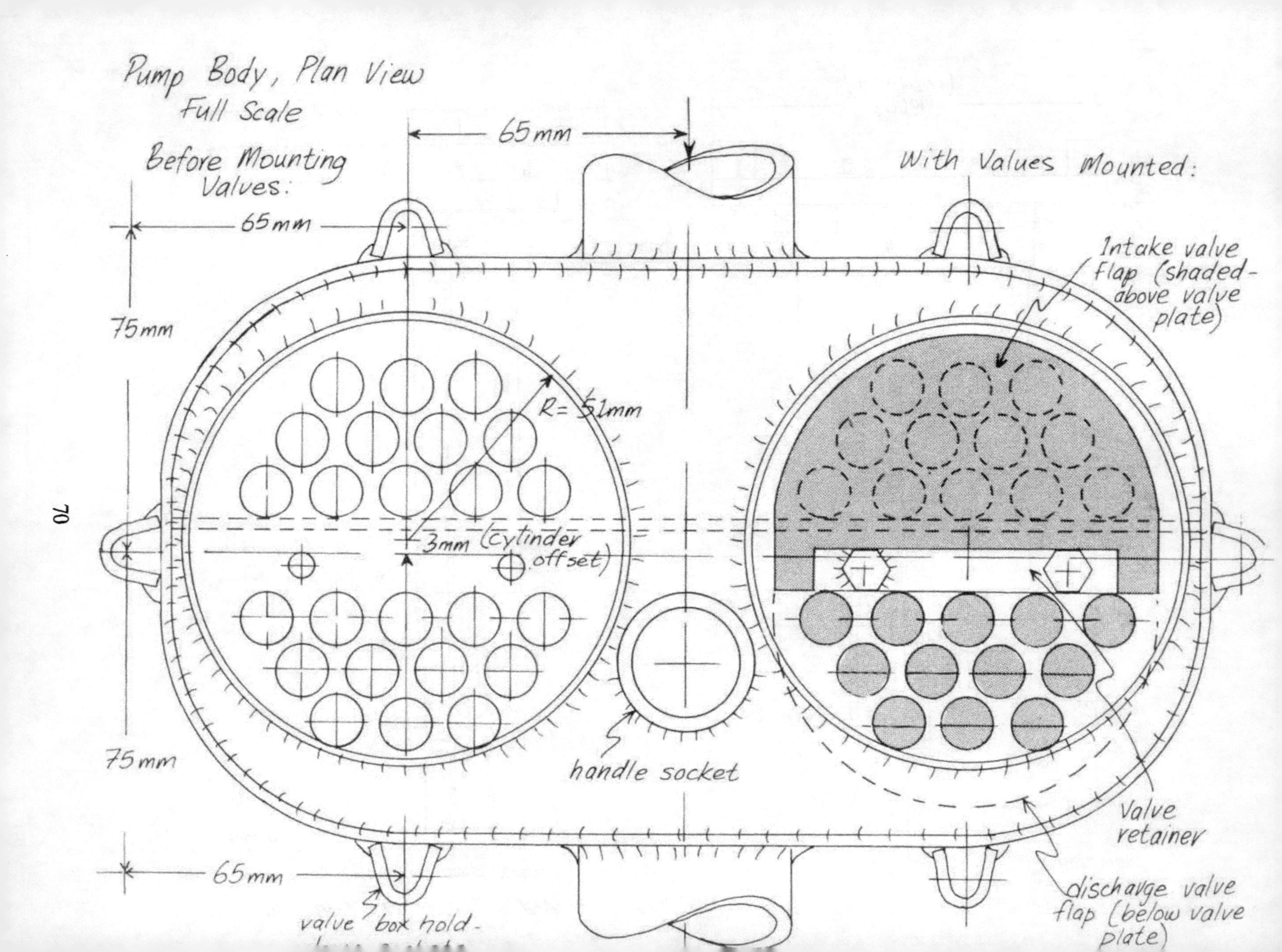
Pump Body, Plan View
Full Scale
Before Mounting Valves:
With Values Mounted:
65mm
65mm
65mm
75mm
75mm
R= 51mm
3mm (cylinder offset)
Intake valve flap (shaded-above valve plate)
handle socket
Valve retainer
dischavge valve flap (below valve plate)
valve box hold-

Pump Body, Cut through cylinder
Full Scale.

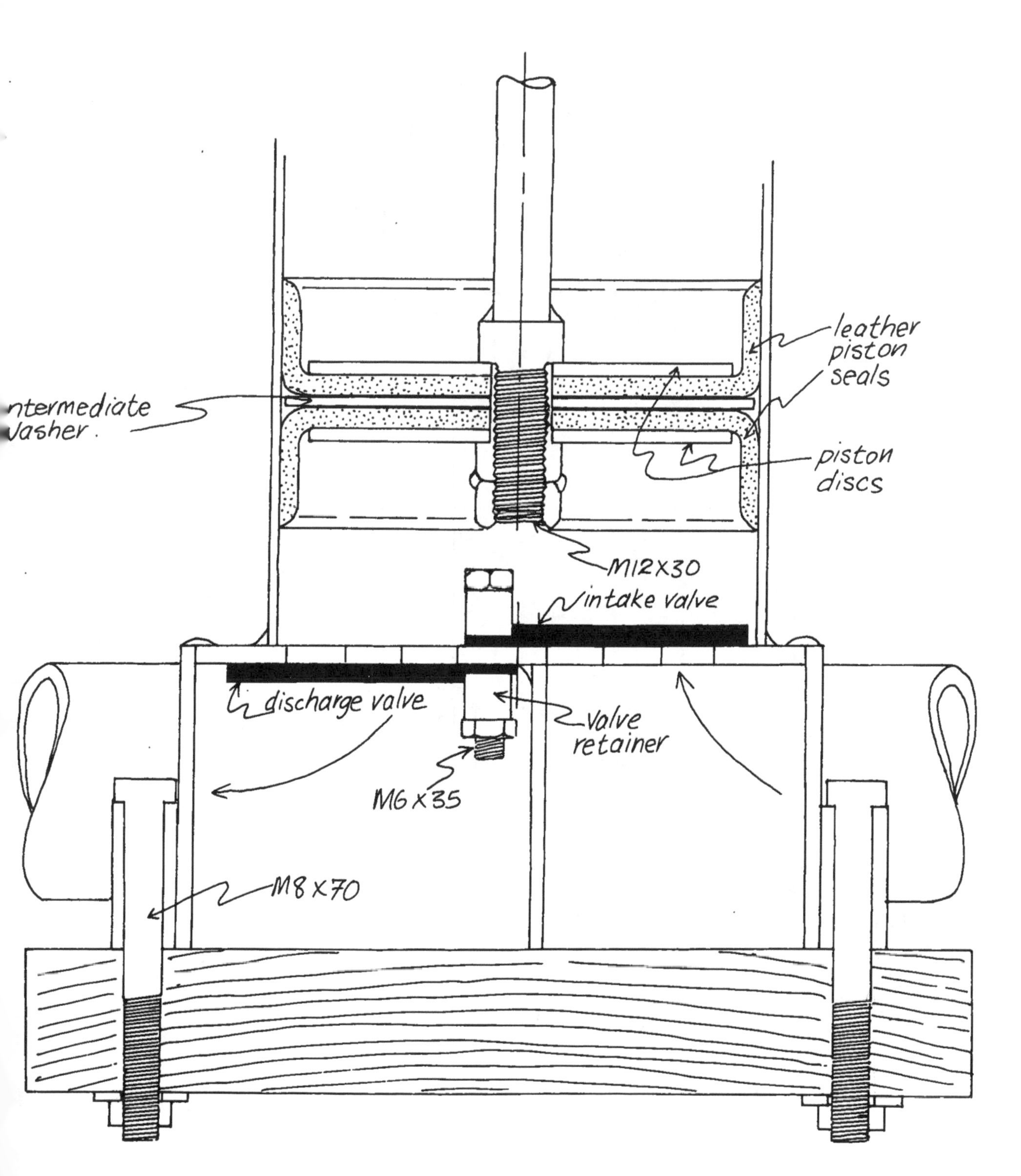

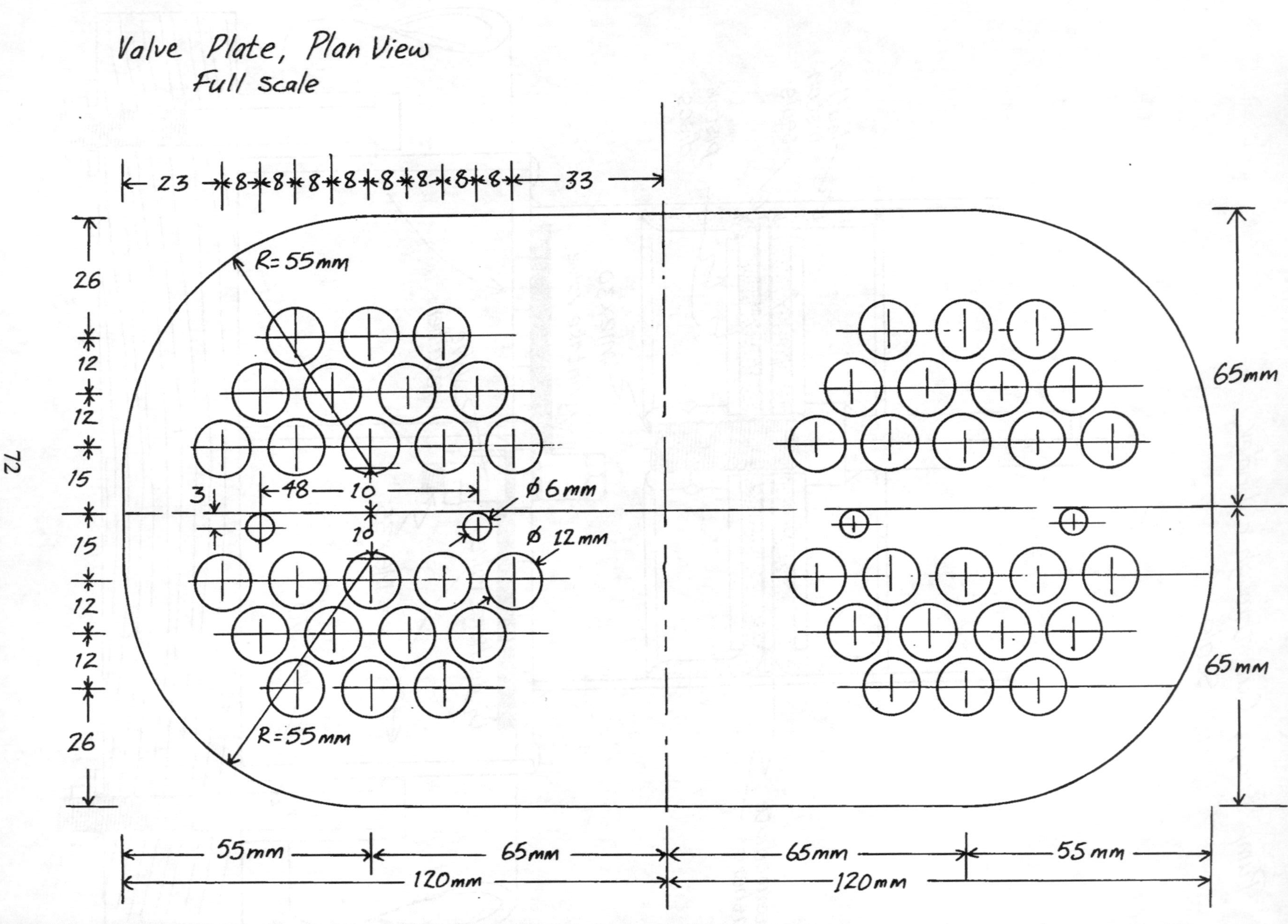
Valve Plate, Plan View
Full scale
23
8
8
8
8
8
8
8
8
33
R=55mm
R=55mm
26
12
12
15
15
12
12
26
3
48
10
10
ø 6mm
ø 12mm
65mm
65mm
55mm
65mm
65mm
55mm
120mm
120mm

Piston Disk, 4 pcs:

NB. Additional 2 disks for use as intermediate washer (see P. 83) ϕ 97–99 mm.

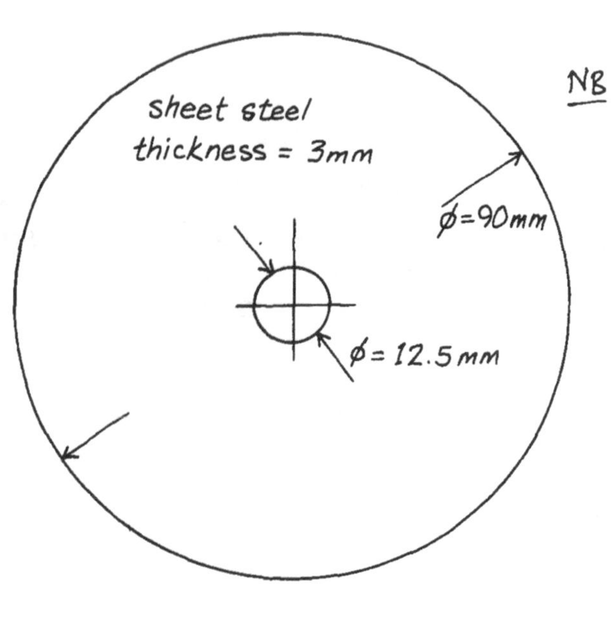

Blank:

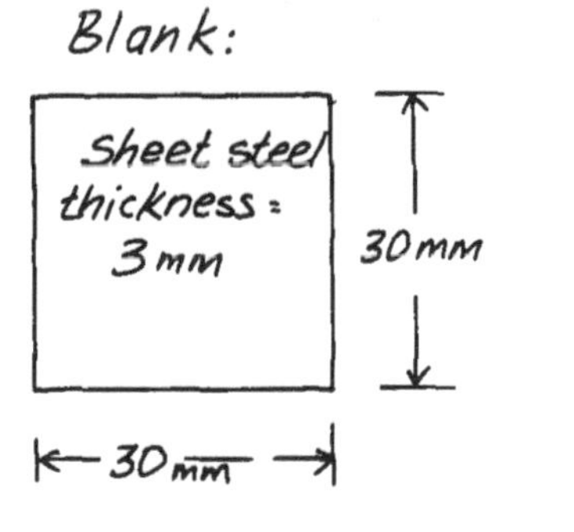


After Bending:

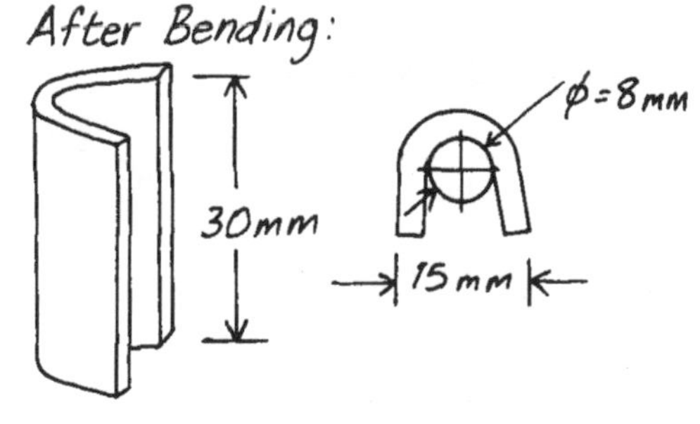


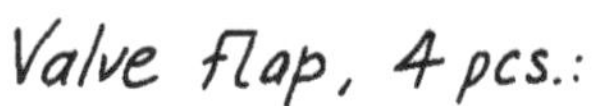


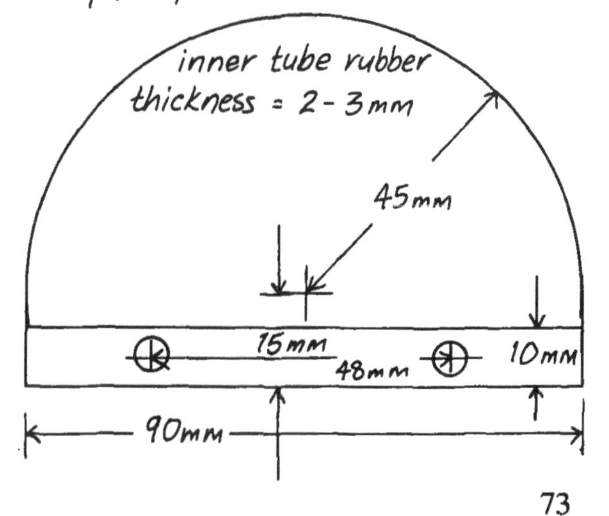


Valve Retainer, 4 pcs.:

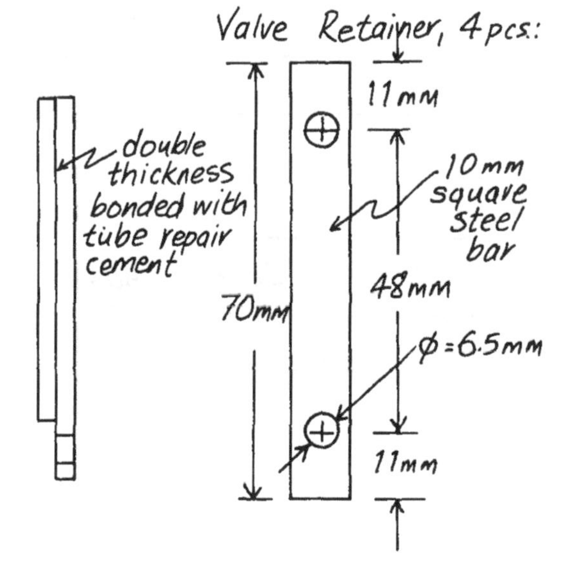

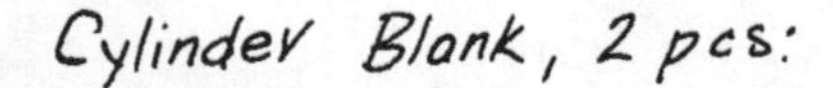


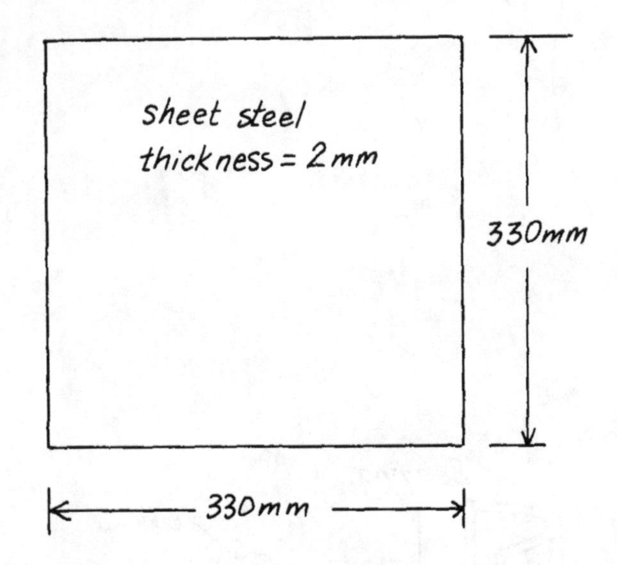


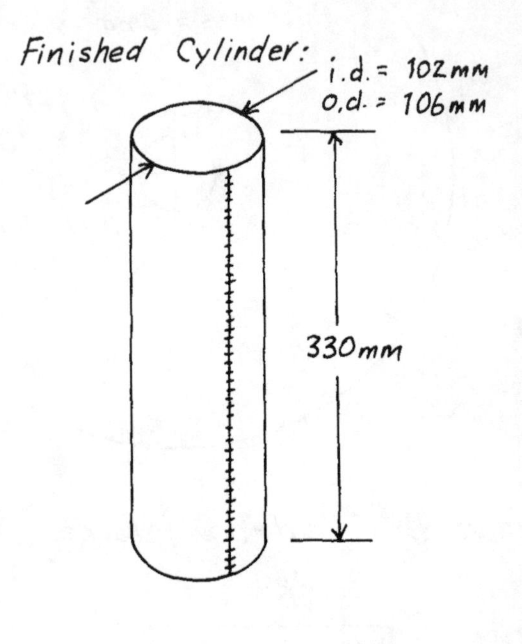


Valve Box Side Band:

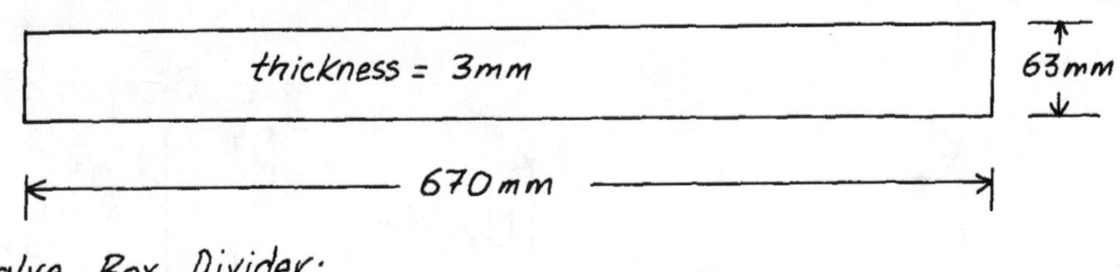


Valve Box Divider:

thickness = 3mm

60mm

240mm

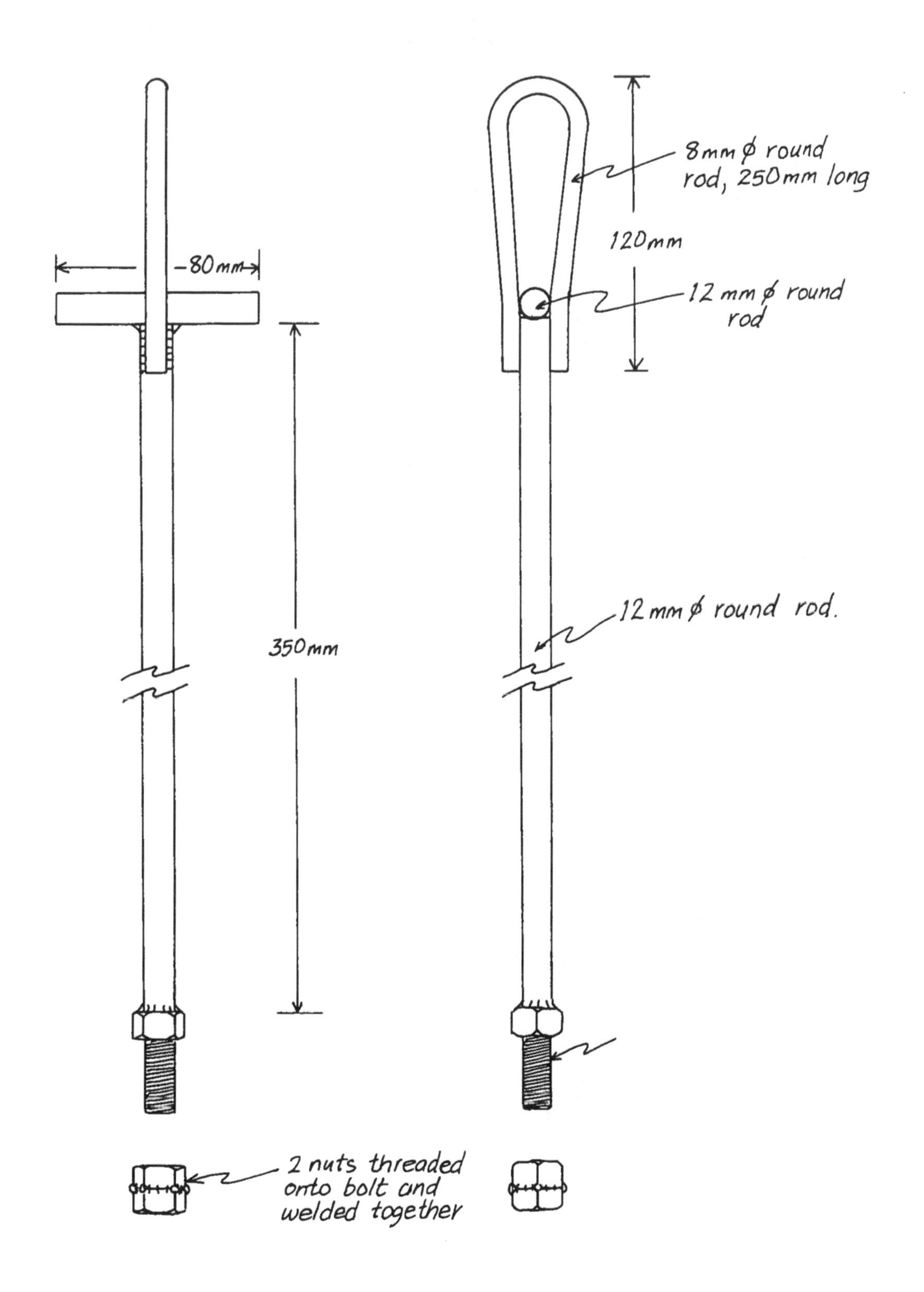
80mm
8mm ø round
rod, 250mm long
120mm
12 mm ø round
rod
350mm
12mm ø round rod.
2 nuts threaded
onto bolt and
welded together

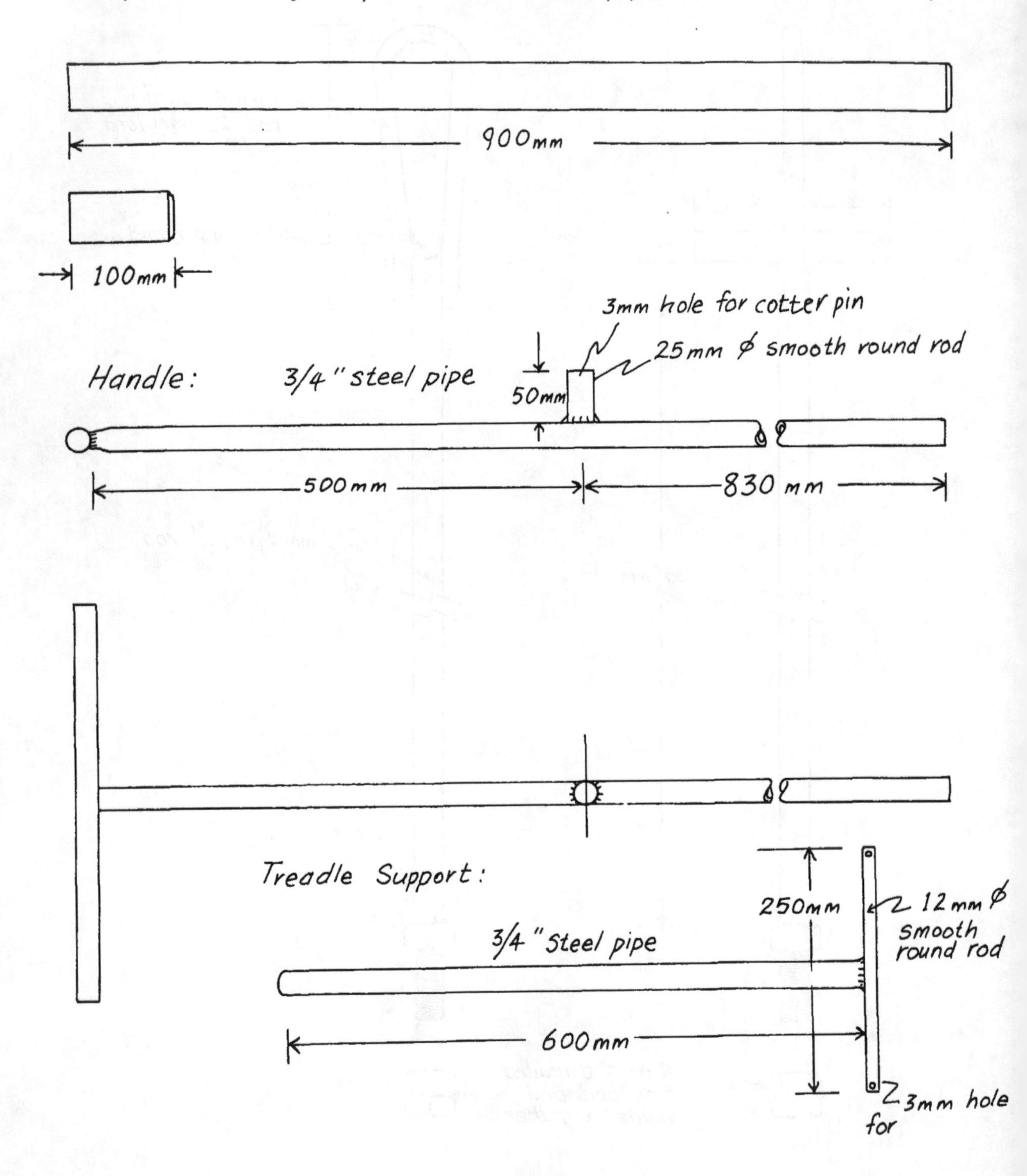
Suction, Discharge Pipes: 1½" steel pipe (49mm o.d., 42mm i.d.).
900mm
100mm
3mm hole for cotter pin
25mm ø smooth round rod
Handle:
3/4" steel pipe
50mm
500mm
830 mm
Treadle Support:
250mm
12mm ø smooth round rod
3/4" Steel pipe
600mm
3mm hole for

# Appendix B: Tool kit components

1. Cylinder rolling tool (5 pieces)
   - o Support bracket
   - o Mandrel
   - o Roller
   - o Vice grip clamps (2)

2. Assembly jig (8 pieces)
   - o Jig and baseplate
   - o Hold-down bar (c/w nuts)
   - o Hole-boring crank and saw-holder
   - o Hole-boring pressure lever
   - o Hole saws (4)

3. Treadle support foot bending form

4. Pump cup forming tools (5 pieces)
   - o Baseplate
   - o Die rings (2)
   - o Wooden moulds (2)

5. Valvc flap hole punch

6. Valve plate boring jig (8 pieces)
   - o Jig plate
   - o Baseplate
   - o Hold-down bar
   - o Pin punch
   - o Bolts (4)

7. Baseboard boring jig

8. Pump cup centre hole punch (optional)

9. Pump cup marking disk (optional)

10. Sample pump (optional)

Printed in the USA
CPSIA information can be obtained
at www.ICGtesting.com
JSHW052019140824
68134JS00027B/2556

9 781853 393129